Automotive Engines Handbook

Automotive Engines Handbook

Todd Stephens

NYRESEARCH PRESS

P R E S S

New York

Published by NY Research Press
118-35 Queens Blvd., Suite 400,
Forest Hills, NY 11375, USA
www.nyresearchpress.com

Automotive Engines Handbook
Todd Stephens

International Standard Book Number: 978-1-63238-854-4 (Hardback)

Cataloging-in-Publication Data

Automotive engines handbook / Todd Stephens.
    p. cm.
Includes bibliographical references and index.
ISBN 978-1-63238-854-4
1. Automobiles--Motors. 2. Automobiles--Power trains. 3. Internal combustion engines.
4. Automobiles--Fuel consumption. I. Stephens, Todd.
TL210 .A88 2022
629.250 4--dc23

Table of Contents

Permissions

Index

It is with great pleasure that I present this book. It has been carefully written after numerous discussions with my peers and other practitioners of the field. I would like to take this opportunity to thank my family and friends who have been extremely supporting at every step in my life.

An automotive engine is a machine which provides the motive power for airplanes and automobiles. It is characterized by a high power to weight ratio that is achieved by using a high rotational speed. There are various kinds of automotive engines such as internal combustion engines, steam engines and electric motors. An internal combustion engine is a motor that produces power by the expansion of gas that is created by the combustion of hydrocarbon gases. Fuels such as diesel, gasoline and ethanol are used by internal combustion engines. Steam engines transform heat into mechanical motion, while the electric motor operates through the interaction between the magnetic field and electric current of a motor in a wire winding to produce force in the form of rotation of the shaft. The various types of automotive engines along with technological progress that have future implications are glanced at in this book. Also included herein is a detailed explanation of the various concepts and applications of these engines. Those in search of information to further their knowledge will be greatly assisted by this book.

The chapters below are organized to facilitate a comprehensive understanding of the subject:

Chapter – Introduction

An automotive engine is a complex, fuel consuming precision-built machine. The different types of automotive engines include diesel engines, inline engines, v-types engines, boxer engines, etc. This chapter discusses in detail these types of automotive engines as well as sources of friction in internal combustion engines.

Chapter – Classification of Automotive Engines

Automotive engines are broadly classified into internal combustion engines, steam engines and electric motors. Internal combustion engines are further divided into gasoline engines and diesel engines. The following chapter provides detailed information about these types of automotive engines.

Chapter – Engine Management Systems and Control Module

The electronic control unit which controls a various actuators on an internal combustion engine to derive optimal engine performance is known as an engine management system. There are various engine management systems such as Trionic and Trionic T5.5 which have been described and thoroughly explained in this chapter.

Chapter – Engine Lubrication Systems

The engine lubrication system is used to distribute oil to the numerous moving parts of an engine for reducing friction between surfaces. Major engine lubrication systems include dry sump systems, wet sump system, total-loss oiling system and oil mist lubrication system. The topics elaborated in this chapter will help in gaining a better perspective about these engine lubrication systems.

Chapter – Technologies for Automotive Engines

Automotive engines make use of wide range of technologies such as air-cooling system, block heaters, radiators, antifreeze, turbochargers, boost controller, tuned exhaust, dual mass flywheel, etc. These technologies associated with automotive engines have been explained in this chapter.

Todd Stephens

1

Introduction

An automotive engine is a complex, fuel consuming precision-built machine. The different types of automotive engines include diesel engines, inline engines, v-types engines, boxer engines, etc. This chapter discusses in detail these types of automotive engines as well as sources of friction in internal combustion engines.

Automotives

The word automotive is not only a word but it also serves as a soul for many human beings. The automotive is one of the greatest creations of the humans. Where it had a drastic development in recent years and a number of new inventions and various technologies have been implemented in it. They are only field where number of new inventions are done day to day. They are one of the important causes for developing jet engines which serves as the one of the greatest creation of this century. Without these jet engines the space travel would have been the dream for the humans and no means of any space research could be done at present.

The automotive serves as a major platform for developing a certain new branches of engineering such as aeronautical and aerospace engineering. They are the combination of both mechanical and electrical engineering. One of the most important part in it is their design. The design plays a vital role because any problems in it may lead to failure of the automotive. At first automotives are created only for the purpose of transport. They are used for travelling from one place to another at earlier periods travelling to longer distances has been impossible but after their creation it has made possible. There are then partially transformed for the purpose of entertainment and after certain period they became one of the part in army where they are used for the purpose of destruction and transportation. Gradually they have been used for the purpose of sports such as racing. The cars and bikes are the most important part of automotives. They are used in large numbers at present due the increased attention from all forms of people. The use of them can be varied from rich persons to poorer persons. They also serve as status symbol for the rich persons. There are number of automotives which are known for its royalty look and status appearance. They serve as most important profitable industry at present. Hence numbers of automotive companies are widely spread among various parts of world. These industries also serve number of employment opportunities to many people. Number of companies holds plenty of shares in this industry. They not only provide employment but also serve as a place for new inventions and creations.

They also involve number of features and steps in it. The most important aspect in designing is the safety, dynamics of vehicle, performance criterion, fuel or economic emissions, cost of production etc. They not only involve car and bikes but also various types in it such as buses, trucks, ships, aero plane, trains etc. They are one of the major reasons for the development of mankind. Thus they have number of advantages in it and they are widely spread. They have one major drawback at present due to increased usage the emission rates has become higher which in turn causes higher degree of pollution. Thus in order to avoid this problems number of new technologies has been developed.

Automotive Engines

Automotive engines are heat engines (internal combustion engines and steam engines) and electric engines used in automotive vehicles. Most automotive engines are piston (reciprocating) internal combustion engines. Automotive engines are divided into four-cycle and two-cycle engines according to the working process and are divided by the method of fuel ignition into spark plug ignition engines (also known as carburetor engines or gasoline engines) and diesel engines, which are self-ignition engines using high-temperature air compressed within the engine cylinders. The fuel mixture of vaporized gasoline and air, blended in the carburetor, is admitted to the cylinders of carburetor piston engines. Piston engines without carburetors also exist; these are equipped with a device for direct injection of fuel into the intake manifold or into the engine cylinder.

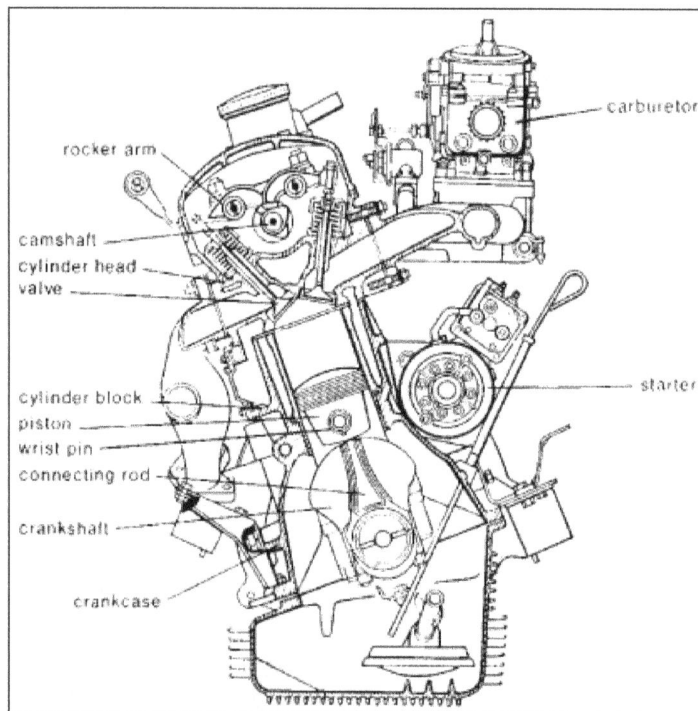

Transverse cross section through the MZMA-412 carburetor engine.

The operating cycle of these engines is the same as that of carburetor engines. In diesel engines, diesel fuel is sprayed through the nozzle of a high-pressure pump directly into the cylinders, where it is mixed with air. Automotive engines are distinguished by the number and placement of the cylinders (in-line engines, V-engines, and so on), by the location of the valves (in the cylinder head or in the cylinder block), by the cylinder capacity, by the motor cooling system (liquid or air), by function, and so on. Four-cycle overhead valve liquid-cooled carburetor-type piston engines are used in modern automobiles, as well as in small and medium trucks. Diesels, which burn a cheaper fuel than gasoline and offer advantages over carburetor type engines in fuel economy and length of service, are used mainly for the propulsion of heavy trucks and large buses. Gasoline motors are superior in ease of design and low initial cost, power per liter of displacement, starting qualities, and control of exhaust smoke, but modern high-speed diesel engines come close to carburetor engines in such important performance characteristics as specific mass (kg/kW or kg/hp), compactness, and noise-free operation. With these advances, and thanks to increases in power per liter of displacement, diesel engines have begun to find applications within the past decade in lightweight trucks and even in passenger automobiles.

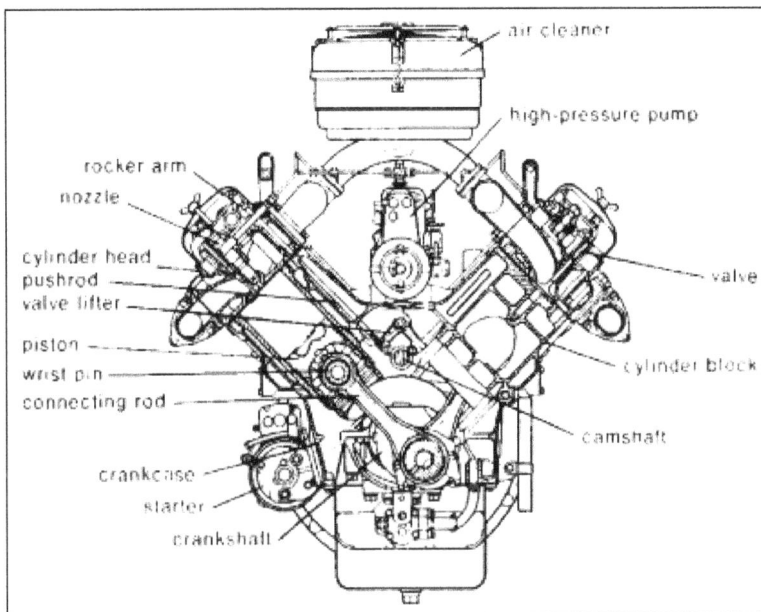

Transverse cross section through the IaMZ-236 four-cycle diesel engine.

Modern four-cycle piston engines consist of a cylinder block usually made integrally with the crankcase, the cylinder head, pistons with sealing rings and oil-control piston rings, connecting rod, crankshaft, flywheel, camshaft, spring-loaded intake and exhaust valves, valve train parts (rocker arm, valve lifter), the drive connecting the crankshaft and camshaft, spark plugs or fuel nozzle, and so on. These engines are equipped with a radiator and fan for the cooling system. They have pumps for forced circulation of lubricant oil and cooling fluid and for bringing fuel from the fuel tank, as well as fuel filters, oil filters, and air cleaners, a starter or cranking motor, ducts for air, gas, fuel, oil, and cooling fluid, and automatic control devices controlling the frequency of rotation of the crankshaft and the temperature of the cooling fluid and fuel mixture.

The power of modern (1968) carburetor piston engines for automobiles ranges from 15 to 310 kilowatts (kW) (20–425 horsepower [hp]), their cylinder displacements from 0.35 to 7.6 liters, compression ratios from 7:1 to 11:1, maximum frequencies of crankshaft rotation from 4,000 to

6,000 rpm, power per liter of displacement from 22 to 50 kW/liter (30 to 70 hp/liter), specific masses from 1.1 to 4 kg/kW (0.8 to 3 kg/hp), and minimum specific fuel rates up to 270 g/(kW ξ hr) [200 g/(hp ξ hr)]; service life to the first major overhaul corresponds to a distance of 75,000–150,000 km or more. In sports and racing piston engines, the crankshaft rotation frequency attains 10,000–12,000 rpm, while the power per liter of displacement sometimes exceeds 150 kW/liter (200 hp/liter). In carburetor piston engines used in heavy-duty trucks, the power is not greater than 220 kW (300 hp); the cylinder displacement ranges from 1.5 to 9.5 liters, the compression ratio from 6.5:1 to 8.5:1, and the maximum frequency of crankshaft rotation 2,500 to 4,000 rpm. Diesel reciprocating engines generate power of 30 to 620 kW (40–850 hp) and feature cylinder displacements of 1.5 to 40 liters, compression ratios of 15:1 to 24:1, maximum frequencies of crankshaft rotation from 2,000 to 5,000 rpm, power per liter of displacement from 11 to 23 kW/liter (15–35 hp/liter), specific masses from 3.4 to 6.8 kg/kW (2.5–5 kg/hp), and minimum specific fuel rates from 205 to 210 g/(kW × hr) [150–155 g/(hp × hr)]; service life to the first major overhaul corresponds to a distance of 150,000 to 300,000 km.

Further development of automotive engines envisages increased power and service life, smaller size, and lower content of noxious exhaust components. The power increase will be achieved primarily through an increase in the frequency of crankshaft rotation in carburetor engines and through supercharging in diesel engines. In addition, the compression ratio is being increased in gasoline engines, and the carburetor will be replaced, to a partial extent, by a system of forced fuel injection. The replacement of conventional piston engines on some lightweight automobiles and light-duty trucks by lightweight and compact rotary-piston engines is highly promising. Once the problem of the fuel economy of gas turbine engines can be solved without exorbitant complications in the design of gas turbine engines, these engines will find broad application at power levels of 750 kW (1,000 hp) and higher. The design of lightweight and compact batteries will make it possible to replace piston engines by electric engines on automotive vehicles traveling within cities.

Types of Automotive Engines

Automotive engines are mainly heat engines or electric engines. The heat engines are classified into internal and external combustion engines or steam engines. Most of the engines are piston or internal and external combustion engines. They are also divided into four-cycle engines and two-cycle engines then by means of working process such as fuel ignition they are divided into gasoline engines, carburetor engines and diesel engines The fuel used in automotive engines can be varied into petrol, diesel, gasoline, Hydrogen etc.

The operating cycles of various types of engines are the same or similar to that of carburetor engines. In case of diesel engines, high pressure of the diesel fuel is sprayed through the tip of the nozzle which is pumped directly to the cylinder, then it is finally mixed with air. Automotive engines are also distinguished by the placement and number of cylinders. Based on it they are classified into in-line engines, V-engines, flat engines, rotary engines etc. Then they are classified on the basis of value location present in the cylinder head or block, cylinder capacity, motor cooling system, function, and so on. In recent times the four-cycles over headed valve and liquid-cooled carburetor-types of piston engines are used in cars and medium and small sized trucks.

Diesel Engines

Diesels are the very low cost fuel which burns cheaper than gasoline and other types of fuels. They also have certain advantages over carburetor type of engines in case of fuel consumption and service length, this type of engine is used mainly for propulsion of heavier trucks and larger buses. Gasoline engines are superior because of their design, lower cost production, power, good starting, and its control over exhaust smoke, but in case of modern higher speed diesel engines has certain properties which makes them come closer towards carburetor engines, the diesel engines nowadays has begun to find applications in various types of automotives. The view of diesel engine is given below.

Straight or Inline Engines

This type of engines has cylinders arranged one after another in a straight line. These engines are considerably easy to build. They also have ultimately low production cost and maintenance cost. They are also very light in weight hence they are mostly preferred in front wheel drive cars. They are extremely fuel efficient than v type engines. The view of straight or inline engine is given below.

V-type Engines

The V-type engines have two sets of cylinders which are placed in 90 degree. It has various advantages such as shorter length, greater rigidity, heavier crankshaft, and low attractive profile. This type of engines has wide applications in sport vehicles mainly in formula1 cars.

Boxer or Flat Engines

Flat engines are another one of the type of engine which has minimum center of gravity than other type of engines, the vehicles which are using this type of engines has certain benefits such as better or good stability and control. These type of engines are wider thus it is difficult to install it in front engine cars.

Friction in Automotive Engines

The current situation of the automotive industry is a challenging one. On one hand, the ongoing trend to more luxury cars brings more and more benefits to the customer and is certainly also an important selling point. The same applies to the increased safety levels modern cars have to provide. However, both of these benefits come with a severe inherent drawback and that is extra weight and, consequently, higher fuel consumption. On the other hand, increased fuel consumption is not only a disadvantage due to the ever rising fuel costs and the corresponding customer demand for more efficient cars. Due to the corresponding greenhouse gas emissions it is also in the focus of the legislation in many countries. Commonly road transport is estimated to cause about 75-89 % of the total CO_2 emissions within the world's transportation sector and for about 20% of the global primary energy consumption. These values do not stay constant; in the time from 1990 to 2005, the required energy for transportation increased by 37% and further increases are expected due to the evolving markets in the developing countries. As industrial emissions decrease, the rising energy demand in the transport sector is expected to be the major problem to achieve a significant greenhouse gas reduction. Consequently, about all major automotive markets introduce increasingly strict emission limits like the national fuel economy program implemented in the CAFE regulations in the US, the EURO regulation in the European Union or the FES in China.

In particular, the European union introduced a limit for the average CO_2 emissions for all cars to be available on the European market of 130 g CO_2/km by 20151. Further, a long term target of 95 g CO_2/km was specified for 2020. To put this into perspective, the average fleet consumption in 2007 was 158 g CO_2/km and it had taken already about 10 years to get down to this value from the 180 g CO_2/km that were achieved in 1998. Now a larger reduction is required in less time. The required reduction of emissions brings also a direct benefit for the customer as the fuel consumption is lowered. It is estimated that the average car consumes every year about 169 litres of fuel only to overcome mechanical friction in the powertrain. There exist several efficient measures to lower the emissions, notably to decrease the weight of the car itself to reduce the energy needed for acceleration, to optimize the combustion process and, of course, to reduce all inherent power losses like the tyre rolling resistance, aerodynamic drag and mechanical losses of the powertrain (combination of engine and transmission) itself.

These measures are, however, not straightforward as some have drawbacks or are even in conflict with each other. For example, smaller cars that offer reduced weight have commonly worse aerodynamic drag as their shape is more cube-like for practical reasons. Also, a reduction in aerodynamic drag brings only a small benefit in the urban traffic due to the low cruising speeds involved. A reduction of tyre rolling resistance is hard to achieve without reduced performance in other areas like handling and traction. Weight reduction is expensive as more and more lightweight and expensive materials have to be used, some of which also require a lot of energy in the production process. In addition, it was shown that there is no positive synergy effect: the combination of several of the mentioned measures reduces their individual efficiency, such that their combination brings less benefit than anticipated.

In contrast, making the powertrain more efficient yields a proportional reduction in CO_2 emissions. For low load operating conditions of Diesel engines friction reduction is even the prime measure to further decrease fuel consumption. While currently a lot of work is done in the automotive industry to reduce the losses caused by the auxiliary systems like the oil or coolant pump, it was shown at hand of a specific engine that the potential for friction reduction in the ICE itself is of comparable magnitude.

Sources of Friction in ICEs

Before any efficient measures to reduce friction in engines can take place, the main friction sources need to be known. At the Virtual Vehicle Competence Center, we use our friction test-rig as shown in figure to investigate the sources of friction for a typical four cylinder gasoline engine; exemplary results for this engine.

Friction measurement test-rig FRIDA during build-up at the Virtual Vehicle Competence Center.
It is shown being applied to an inline four cylinder gasoline engine with 1.8 litres total displacement.

The chart confirms the commonly propagated main sources of friction: the piston-liner contact is the cause for about 60% of the total mechanical losses, while the journal bearings in the crank train (main and big end bearings) contribute about 30%. Finally, the valve train generally represents the third main source of friction and typically causes losses that equal roughly about the half of the power losses in the journal bearings.

While not only the amount of friction is different between the various sources, also the character of friction, namely the lubrication regime itself, is also notably different. While the journal bearings are generally full film lubricated with no metal metal contact occurring under normal operating conditions, parts of the piston assembly experience metal-metal contact under high load. In particular, the top ring of the piston has metal-metal contact every time it passes the top dead center as no oil can reach this point. This is of particular severity as at firing top dead center a large force acts on the top ring and presses it onto the cylinder liner during the downward motion of the piston. Besides the fact that the piston assembly has generally been the largest contributor to the total mechanical losses, it has several other important functions. Amongst others it has to seal the combustion chamber in both directions, both to avoid so-called blow-by gases from entering the engine housing, as well as to control the amount of lubricant being left on the cylinder liner. The blow-by gases have to be controlled as these both cause a loss in convertible energy by decreasing the available cylinder pressure as well as have a negative deteriorating impact on the lubricant properties. The oil being left on the cylinder liner needs to be carefully controlled as well: while a certain amount of oil is necessary to provide sufficient lubrication for the piston rings, it is burned during combustion. Burning too much oil needs to be avoided not only for practical reasons as it needs to be replaced (increased service demand), but also as some of its contents are problematic for the exhaust aftertreatment systems.

Additionally, depending on operating condition unstable behaviour of the piston rings may occur like ring flutter (rapid oscillating movement of the piston ring in its groove) or ring collapse (inward forces on the ring exceed the ring tension), which needs to be avoided in practical designs. To summarize, the piston assembly has to fulfil many functions. For focusing solely on friction it is, therefore, not used in this work.

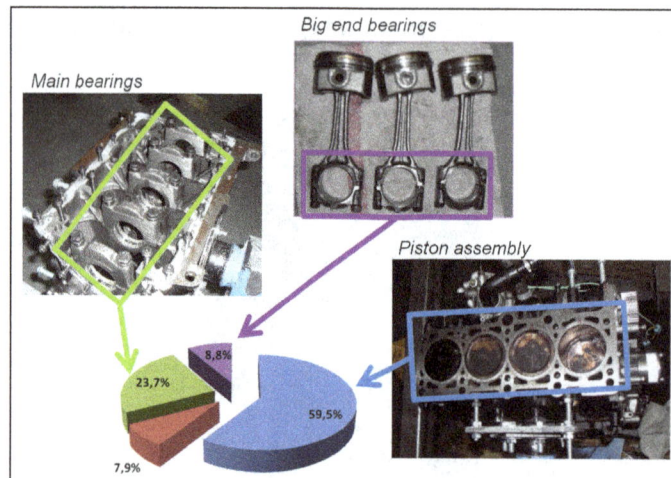

Examplary relative contributions to the total friction losses in an inline 4 cylinder gasoline engine with 1.8 litre total displacement. Shown in blue is the contribution of the piston assembly, in green the contribution of the main bearings, in violet the amount caused by the big end bearings and, finally, the red part shows the contribution of all other components like seals etc. The valve train is not included in these results, also all auxiliary systems (oil pump etc.) are removed.

Calculating Power Losses due to Friction in the Journal Bearings

In contrast to the piston assembly that has to perform a large number of tasks which are partially conflicting as previously discussed, journal bearings are due to their apparent simplicity particularly suited to discuss the sources of friction.

Journal bearings are from their appearance simple devices; generally formed from sheet metal they are typically low cost parts, with one bearing shell costing a few single Euros or less. However, this simplicity is misleading, as in fact they have to combine a wide range of properties which impose conflicting requirements on the material properties to be used. While the bearing material should be hard to resist wear, in the engine it shall also embed well debris particles that originate from wear or even from the original manufacturing process of the engine housing. For the latter property softer materials are beneficial which conflicts with the requirement to resist wear. These requirements led to the development of multi-layer bearings, where each layer is optimized for a specific task.

In the following a method is described that accounts for many of the essential physical processes that occur in journal bearings during operation and allows to accurately predict the power losses due to friction. The method is developed while discussing these processes and its validity is shown by numerous comparisons to experimental data.

While the focus in the following is on monograde oils as they are used in large stationary engines, the results also apply correspondingly to multigrade oils with their shear rate dependence taken into account.

In the following the results from a number of works are presented in a shortened form with a particular focus on the results and their context.

An Isothermal EHD Approach

In an ICE, journal bearings are generally exposed to different operation conditions in terms of load, speed and temperature. As depicted in figure, depending on relative speed, load and viscosity the operating conditions reflected as friction coefficient may range from purely hydrodynamic lubrication with a sufficiently thick oil film to mixed or even boundary lubrication with severe amounts of metal to metal contact.

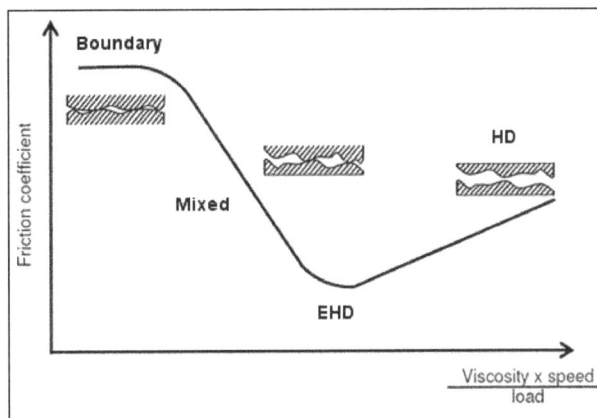

The Stribeck-plot showing the different regimes of lubrication: hydrodynamic (HD), elastohydrodynamic (EHD), mixed and boundary lubrication.

To calculate the movement of the journal under the applied load and the corresponding pressure distribution within the oil film an average Reynolds equation is used, that takes into account the roughness of the adjacent surfaces. When the typical minimum oil film thickness is of comparable magnitude to the surface roughness, the lubricating fluid flow is also affected by the surface asperities and their orientation. To account for this modification of the fluid flow we use the average Reynolds equation as developed by Patir and Cheng, which can be written in a bearing shell fixed coordinate system as:

$$-\frac{\partial}{\partial x}\left(\theta\phi x\frac{h^3}{12\eta_p}\frac{\partial p}{\partial z}\right)-\frac{\partial}{\partial z}\left(\theta\phi z\frac{h^3}{12\eta_p}\frac{\partial p}{\partial z}\right)$$
$$+\frac{\partial}{\partial x}\left(\theta(\bar{h}+\sigma_s\phi_s)\frac{U}{2}\right)+\frac{\partial}{\partial t}\left(\theta\bar{h}\right)=0,$$

where x, z denote the circumferential and axial directions, θ the oil filling factor and h, $\bar{h}$ the nominal and average oil film thickness, respectively. Further, U denotes the journal circumferential speed, ηp the pressure dependent oil viscosity and σs the combined (root mean square) surface roughness. ϕx, ϕz, ϕs represent the flow factors that actually take into account the influence of the surface roughness.

To describe mixed lubrication another process needs to be taken into account, namely the load carried by the surface asperities when metal-metal contact occurs.

The corresponding quantity is the asperity contact pressure pa and together with the area experiencing metal-metal contact, Aa, and the boundary friction coefficient µBound these yield the friction force RBound caused by asperity contact,

$$R_{Bound}=\mu_{Bound}\cdot p_a\cdot A_a.$$

To describe the metal-metal contact we use the Greenwood and Tripp approach , that is shortly outlined in the following.

The theory of Greenwood and Tripp is based on the contact of two nominally flat, random rough surfaces. The asperity contact pressure p_a is the product of the elastic factor K with a form function $F_{\frac{5}{2}}$ F (Hs),

$$p_a=KE*F_{\frac{5}{2}}(H_s),$$

where H_s is a dimensionless clearance parameter, defined as $H_s=\dfrac{h-\bar{\delta}_s}{\sigma_s}$, with σ_s being the combined asperity summit roughness, which is calculated according to,

$$\sigma s=\sqrt{\sigma^2_{s,J}+\sigma^2_{s,S}}$$

and δs being the combined mean summit height, $\bar{\delta}_s=\bar{\delta}_{s,J}+\bar{\delta}_{s,S}$, where the additional subscript J and S denotes the corresponding quantities of the journal and the bearing shell, respectively.

Further, E^* denotes the composite elastic modulus, $E^* = \left(\dfrac{1-v_1^2}{E_1} + \dfrac{1-v_2^2}{E_2} \right)^{-1}$, where v_i and E_i are the Poisson ratio and Young's modulus of the adjacent surfaces, respectively. The form function is defined as:

$$F_{\frac{5}{2}}(H_s) = 4.4086.10^{-5}(4-H_s)^{6.804} \text{ for } H_s < 4$$

$$= 0 \text{ for } H_s \geq 4,$$

which shows that friction due to asperity contact sets in only for $H_s < 4$ and further sensibly depends on the minimum oil film thickness as this quantity enters $= 0$ for $H_s \geq 4$, with almost 7th power.

For the calculation of the Greenwood/Tripp parameters a 2D-profilometer trace was used that was performed on an run-in part of the bearing shell along the axial direction.

Modern engine oils include friction modifying additives like zinc dialkyl dithiophosphate (ZDTP) or Molybdenum based compounds to lower friction and wear in case metal-metal contact occurs. For the Greenwood and Tripp contact model we employed in the following a boundary friction coefficient of $\mu_{Bound} = 0.02$.

The different contributions to friction, as listed in equation. $\dfrac{\partial}{\partial x}\left(\theta \phi x \dfrac{h^3}{12\eta_p} \dfrac{\partial p}{\partial z} \right) \cdot + \dfrac{\partial}{\partial x}\left(\theta(\overline{h}+\sigma_s \phi_s)\dfrac{U}{2} \right) + \dfrac{\partial}{\partial t}\left(\theta\overline{h} \right) = 0,$ and $p_a = KE * F_{\frac{5}{2}}(H_s),$ are generally not independent from each other. A reduction in lubricant viscosity, while decreasing hydrodynamic losses, may cause - depending on the load - an overly increase in asperity contact as the oil film thickness enters equation $F_{\frac{5}{2}}(H_s) = 4.4086.10^{-5}(4-H_s)^{6.804} \text{ for } H_s < 4$

$$= 0 \text{ for } H_s \geq 4, \text{ with almost 7th power.}$$

Testing Method

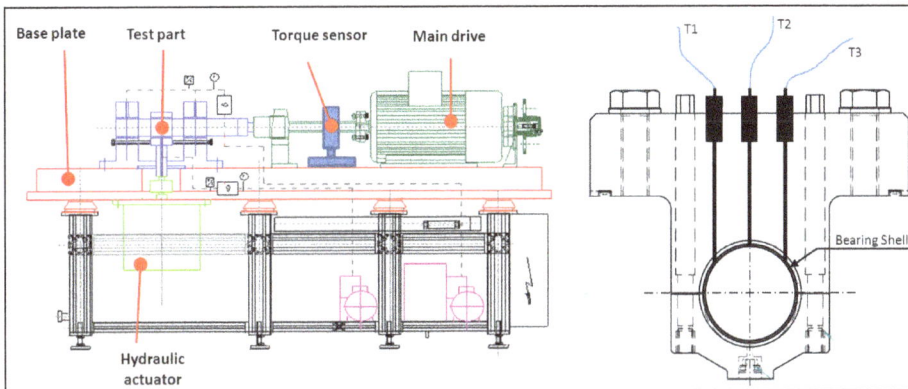

In the above figure, left: schematic drawing of the journal bearing test rig LP06: test part denotes the location of the test bearing, torque sensor the HBM T10F sensor used for friction moment measurement. Right: Drawing of the test con rod with test bearing showing the location of the temperatures sensors: T2 sits in the center at 0° circumferential angle, with T_1 and T_3 at ±45° circumferential angle, respectively.

MIBA[2]'s journal bearing test rig LP06 was used for the experimental measurements. It is sketched in figure and consists of a heavy, elastically mounted base plate which carries the two support blocks, the test con rod with the hydraulic actuator and the driveshaft attached to the electric drive mechanism. The hydraulic actuator applies the load along the vertical direction, which is consequently defined as 0° circumferential angle.

The friction torques arising from all three journal bearings were measured at the driveshaft; for the comparisons load cycle averaged values of the friction moment are used.

The LP06 is equipped with a number of temperature sensors to capture the occurring temperatures at various points of the test rig; to this task temperature is measured by using thermocouple elements of type K that have an accuracy of ±1 °C. Besides two temperatures in the con rod and the oil outflow temperature, the bearing shell temperatures of the test and support bearings are measured at three different points at the back of each corresponding bearing shell. As shown in figure, two of these temperature sensors are located at ±45° circumferential angle from the vertical axis and the third in the middle at 0° circumferential angle.

For the bearing tests following conditions were maintained: for test- and support-bearings steel-supported leaded bronce trimetal bearings with a sputter overlay were employed; for each test-run new bearings with an inner diameter of 76 mm and a width of 34 mm were used and mounted into the test rig with a nominal clearance of 0.04 mm (10/00 relative clearance). A hydraulic attenuator applied the transient loads with the corresponding peak loads of either 41 MPa, 54 MPa, 70 MPa or 76 MPa. For a convenient comparison of the results to other works the peak load is expressed in MPa to account for the involved bearing dimensions. This is conducted by dividing the load force by the projected bearing area (product of bearing width and bearing diameter). Therefore, the peak loads of 106 kN and 180 kN correspond to 41 MPa or 70 MPa, respectively, for the present bearing dimensions. In the following, the corresponding peak loads are used to distinguish between the different transient load cases.

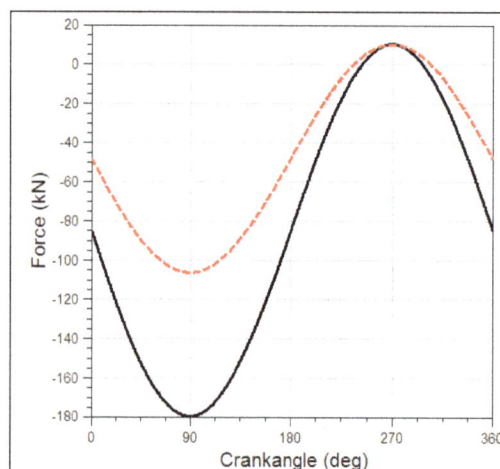

In above figure, plot of the loads applied to the test bearing: at a frequency of 50 Hz a sinusoidal load is applied along the vertical direction with a preload of -10kN and a peak load of either 180 kN for the 70 MPa load case (shown as solid black line) or a peak load of 106 kN for the 41 MPa load case (red dashed line).

The different oils were preconditioned to 80±5 °C inflow temperature. After the test-run the wear at several points in the journal bearings was measured and the so obtained wear profiles were included in the simulation model.

Simulation

For the simulation a model of the LP06 was setup within an elastic multi-body dynamics solver (AVL-Excite Powerunit[3]). The simulation model consists of the test con rod including the test bearing, the two support-blocks with journal bearings and the shaft running freely, but supported by the adjacent bearings. All structure parts are modeled as dynamically condensed finite element (FE)-structures.

The three journal bearings, 76mm in diameter and 34mm width, are represented as EHD or TE-HD-joints, respectively.

To obtain realistic dynamic lubricant viscosities for the calculations, the viscosities and densities of fresh SAE10/SAE20/SAE30 and SAE40 monograde oils were measured at different temperatures in the OMV-laboratory[4]. To obtain a pressure dependent oil-model for the simulation, the pressure dependency was impressed onto the measured viscosities by applying the well known Barus-equation with the coefficients from. The so resulting dynamic viscosities correspond qualitatively to experimental data. Further, a dependence on hydrodynamic pressure was impressed onto the lubricant density following the data found experimentally by Bair et al.

The dynamic viscosities and oil-densities are shown for the SAE10, SAE20, SAE30 and SAE40-oils in figure. As can be seen in these figures, a hydrodynamic pressure of about 60 MPa leads to roughly a doubling of the dynamic viscosity and, therefore, to a strong increase in the related hydrodynamic losses. While for now the presented calculations do not take into account the local temperatures of the lubricant in the bearing itself, the strong variation of the physical properties of the oil with temperature show the importance of defining a representative global lubricant temperature.

Deriving the Oil-Temperature

A plausible choice of this temperature is important as it directly relates to the lubricant viscosity and consequently acts on the minimum oil film thickness and the amount of asperity contact.

For the presented pressure dependent lubricant model, the calculation of the global oil temperature is straightforward: as the oil viscosity increases strongly for hydrodynamic pressures exceeding about 1 MPa, the hydrodynamic losses in the lubricant are expected to be dominated by this thickening in the high-load area of the bearing. This argument is also supported later on by the simulation results which predict hydrodynamic pressures of up to 120 MPa in large areas in the bearing. Following this line of argument, the global oil temperature is estimated from the measured bearing back temperatures, by averaging the test and support bearing back temperatures that are located at ±45° circumferential angle (T_1, T_3) of the one in the high load zone, T_2, as shown in figure. Although the so obtained temperature is rather high in comparison to the oil inflow

Temperature it is expected to realistically estimate the hydrodynamic losses as well as the amount of asperity contact, as this temperature describes closely the oil viscosity in the high load zone.

The such calculated oil-temperatures are depicted in table for the load cases studied in the following and for simplicity the same oil temperature is used for all three bearings.

Surface Profiles

For a sufficiently accurate calculation of the asperity contact, it is necessary to use realistic surface shapes in the simulation. Ideal geometric shapes are not suitable for this task, as due to elastic deformation of the structure under load, the bearing pin would express overly large pressures on the outermost nodes of the bearing shell, leading to unrealistically high amounts of asperity contact. This in turn causes an overestimation of the friction moment.

Therefore, the bearing shell surface of the test bearing was measured for wear at several points after the test runs. The wear data obtained from the two performed SAE10-oil test-runs were averaged and symmetrized as we do not include misalignments due to imperfect mounting in the simulation. The such obtained wear profiles, depicted in figure were used for all consequent calculations (all lubricants and loads) as it was found that the results with individual wear profiles prepared for every lubricant caused only negligible differences in comparison to the results obtained with the SAE10-wear profile. Similar wear profiles were used for the support bearings, which, however, are less critical as each of these carries only half of the total load.

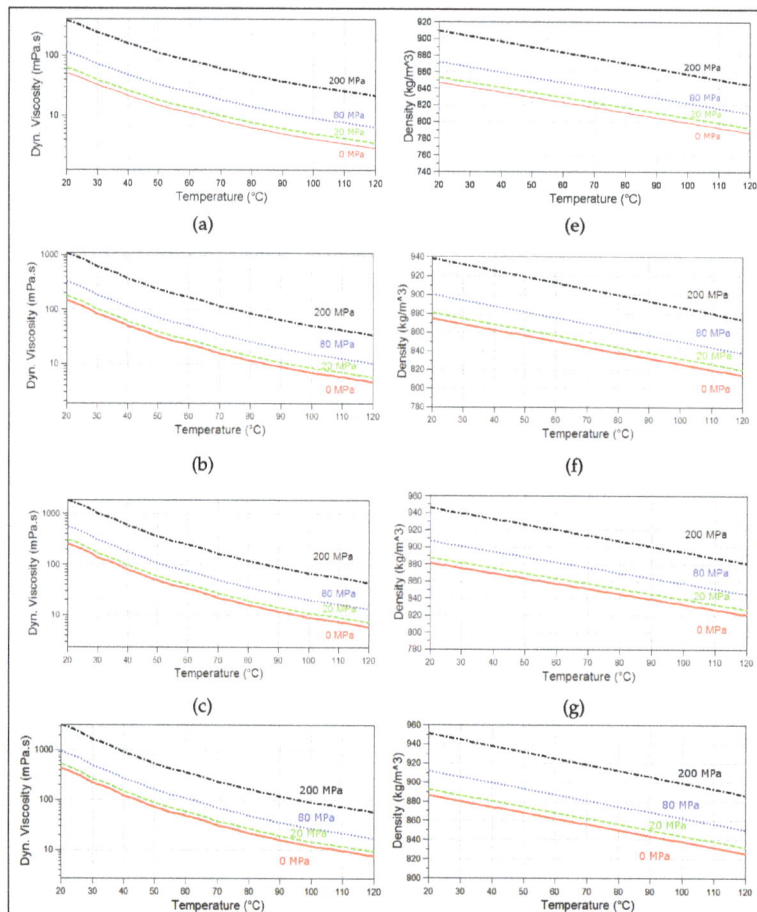

Dynamic viscosities (a-d) and densities (e-h) of the lubricants as described by the oil model for SAE10/SAE20/SAE30/SAE40 (top to bottom): the solid red lines denote the corresponding physical property at 0 Pa, the green dashed lines at 20 MPa, the blue dotted lines for 80 MPa and the black dash-dotted lines for 200 MPa.

Table: Calculated oil-temperatures for the studied load cases (denoted as subscript) and the different oils.

SAE10	$T_{41MPa}[^{o}C]$	$T_{70MPa}[^{o}C]$
	87.4	94.6
SAE20	$T_{41MPa}[^{o}C]$	$T_{70MPa}[^{o}C]$
	89.2	96.8
SAE30	$T_{41MPa}[^{o}C]$	$T_{70MPa}[^{o}C]$
	89.8	97.9
SAE40	$T_{41MPa}[^{o}C]$	$T_{70MPa}[^{o}C]$
	91.8	99.6

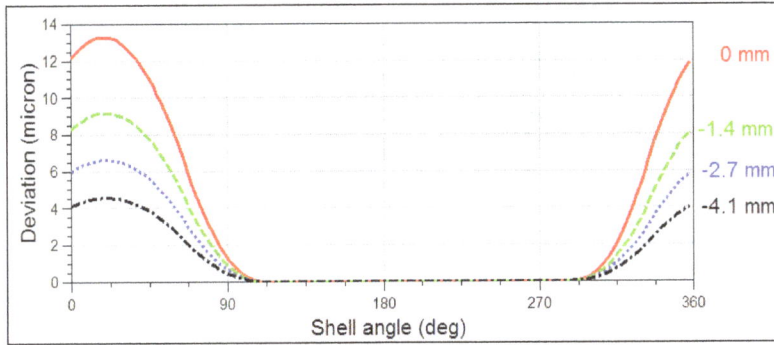

Surface profile used for the test bearing in the simulation, shown as deviation (in μm) from the nominal geometrical shape of the bearing shell for different axial cuts: the red line denotes the deviation at the outermost bearing nodes, 0mm from the shell edges, the green dashed line the deviation −1.4mm from the shell edges, the blue dotted line the deviation −2.7mm from the bearing edges and the black dash-dotted line shows the deviation −4.1mm from the bearing edges. The deviations in the middle of the journal bearing, along axial direction from −4.1mm to −17mm from the bearing edges, stay rather constant and are almost identical to the deviation for −4.1mm away from the bearing edges.

Results

In the following, the results obtained from simulation are compared to the experimental results. For this task, the results are discussed starting from full fluid film lubrication (purely hydrodynamic losses), as it is the case for SAE40, to working conditions which progress increasingly into mixed lubrication, like it is the case for SAE10, where friction power losses due to metal-metal contact become significant.

The simulations for the different lubricants were conducted and the calculated friction moments are shown and compared to the experimental values in figure. The obtained experimental values range from as low as 3 to 8 Nm, where the individual average friction torques are measured (with the exception of the 3000rpm cases where only a smaller amount of measurement data is available) with an accuracy of about 0.5 Nm (denoted as black bars in the plot).

Regarding the general trends, the resulting average friction moment scales with the applied load and for a given load, with the journal speed. For a specific load/journal speed case, the friction moment depends significantly on the different lubricant viscosities.

Due to the large number of cases shown in figure the focus will be in particular on the results for a journal speed of 2000rpm and two different load cases in the following.

Comparing the results predicted by simulation to the experimental data, we start by looking first at the SAE40 lubricant. From experience it is known that for a load of 41 MPa SAE40 oil provides sufficient lubrication for reliable long term use and, thus, avoids asperity contact. This is confirmed by the results from simulation, as these do not show any asperity contact. The same is almost true for the 70 MPa case, although simulation predicts about 0.5 W power losses due to the metal-metal contact in this case. However, this amount is negligible compared to the total losses of 626 W.

Starting from the purely hydrodynamic case with SAE40 lubrication, excluding one case all results from simulation are within the measurement uncertainty of the experimental results and commonly even closer to the average friction torque. The result calculated for the excluded case is also close to the measurement uncertainty.

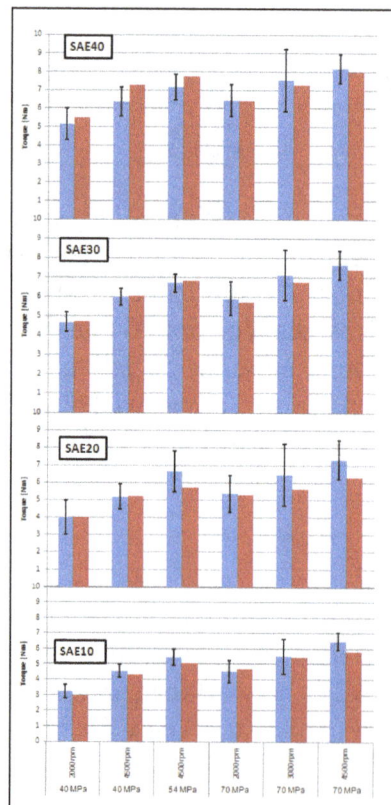

Comparison of the friction torques measured on the test-rig (blue) together with the measurement uncertainty (black bars) with the results from simulation (illustrated in red) for different journal speeds (2000/3000/4500rpm), different dynamic loads (40/54/70 MPa) and different lubricants (SAE40/SAE30/SAE20/SAE10).

Lowering the lubricant viscosity by about 25% compared to the SAE40 lubricant, one arrives at the cases with lubrication with SAE30. Here, the onset of mixed lubrication occurs for the highest dynamic load of 70 MPa at the lowest journal speed of 2000rpm. For this smaller load, there is still no asperity contact predicted and all losses are caused by the dynamic viscosity of the lubricant. Beginning with this lubricant class the validity of the chosen asperity contact model starts to get tested. While the power losses are still strongly dominated by hydrodynamic losses, the power losses due asperity contact contribute about 1% (7 W averaged over a full crank cycle) to the total power losses. The corresponding asperity contact pressure, however, increases for the SAE30 lubricant to a maximum of 15 MPa, which indicates that wear starts to occur in the highly loaded parts of the journal bearing.

However, comparing the results from simulation with the experimental data, the same level of accuracy can be seen in figure. In fact, the actual predicted value from simulation is generally even closer to the measured averaged value.

Lowering the lubricant viscosity once more to SAE20, the journal bearing experiences more and more mixed lubrication. While metal-metal contact is still absent for a load of 41 MPa, it starts to become significant for the 70 MPa load case. In total the losses due to metal-metal contact are still rather low representing only about 4% of the total power losses for the worst case as shown in figure. Due to the reduced viscosity in comparison to SAE30-oil, the experimentally observed average friction moments are reduced, however, asperity contact begins to reduce the benefit of the decreased hydrodynamic losses. As can be seen from the results, also for this case the prediction accuracy of the presented simulation method lies for all cases studied within the measurement uncertainty of the experimental data.

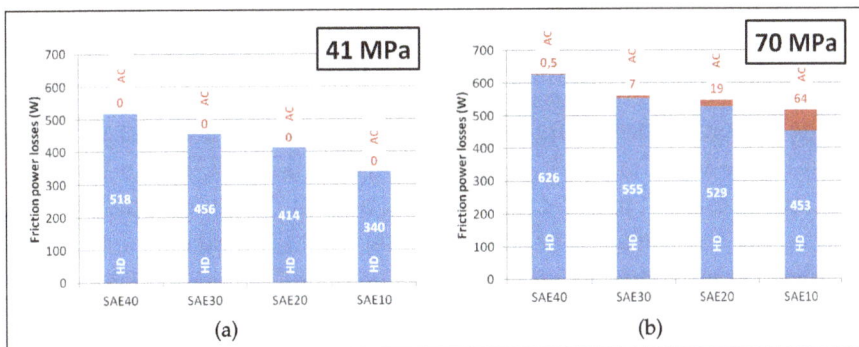

Comparison of the contributions to the friction power losses in the test bearing for a load of (a) 41 MPa and (b) 70 MPa for 2000rpm journal speed calculated for the different oils: hydrodynamic losses are denoted as HD and losses due to asperity contact are denoted as AC.

Finally, arriving at the lowest viscosity lubricant SAE10 this case represents the most severe operating conditions in terms of mixed lubrication that are investigated in this work. For the most severe case of the highest dynamic load of 70 MPa and the lowest journal speed studied of 2000rpm, already about 14% of the total power losses are caused by metal-metal contact. While the journal bearing might endure these conditions on the test-rig still for a rather long time, the amount of metal-metal contact is much to high to be allowed in real world applications. Significant wear occurs every load cycle which does not stabilize after a run-in period and leads, therefore, to a constant wear of the bearing shell. For the lower load case of 41 MPa, simulation still predicts no asperity contact and attributes all friction power losses to the hydrodynamic losses.

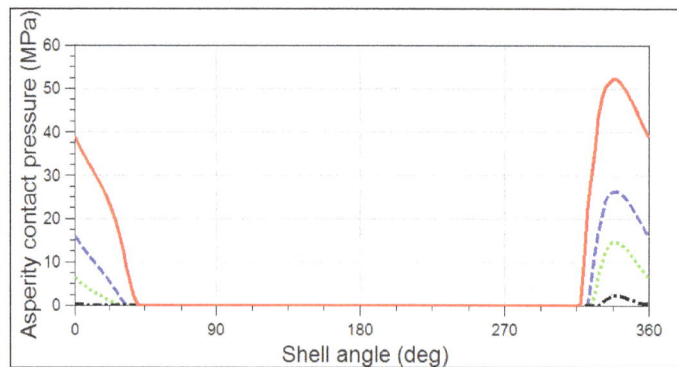

Comparison of the asperity contact pressures occurring at the outermost bearing shell edges for a load of 70 MPa for the different oils: SAE10 denoted as solid red line, SAE20 denoted as dashed blue line, SAE30 and SAE40 shown as green dotted line and as black dash-dotted line, respectively.

Finally, the ability of the presented method to predict the existence of metal-metal contact is put to test. For this task, a bearing durability test is investigated. For this test, the operating conditions are made even more severe by increasing the dynamic load to a maximum of 76 MPa at a journal speed of 3000rpm and increasing the oil inflow temperature of the SAE10 lubricant to 110 °C. In comparison to the previous operating conditions with an oil inflow temperature of 80°C, this temperature increase causes the lubricant viscosity to decrease by more than 50%. These operating conditions lead consequently to bearing shell temperatures exceeding 130 °C. As significant metal-metal contact occurs for these operating conditions, it can be detected by contact voltage measurements. For this measurement, a voltage is applied e.g. in form of a charged capacitor between the journal and the bearing. As the lubricant has only a poor electrical conductivity, the capacitor stays charged and the voltage remains unchanged. When metal-metal occurs, the capacitor can discharge due to the corresponding increased electrical conductivity; this process can be observed as change (decrease) of the voltage.

A comparison of the experimental contact voltage measurement and the predicted metal-metal contact is shown in figure together with the applied dynamic load. It can be seen that when the load exceeds a certain threshold, metal-metal contact occurs. When compared to the results from simulation, the onset and the duration of the calculated metal metal contact agrees very well with the measured contact voltage data.

Overall, the presented simulation method appears to describe the actual processes in the journal bearing sufficiently well, as it predicts the friction moment accurately and reliably over a large range of working conditions, which range from purely hydrodynamic to significantly mixed lubrication.

Other important properties related to reliability in lubricated journal bearings are the peak oil film pressure (POFP) and the minimum oil film thickness (MOFT) , that are depicted in figures for the investigated lubricants.

As shown in figure, the POFPs change significantly from about 90 MPa to 120 MPa between the two different loads, but do not vary significantly between the different lubricants at the same load.

For a load of 41 MPa the results show that the MOFT is for all investigated oil-classes above 1.5 μm, which is the asperity contact threshold. Therefore, no metal-metal contact occurs, which can also be seen in the power losses shown in figure.

Further, it is interesting to note that the MOFT decreases by about 0.5 μm for every decrease in SAE-class; while the MOFT is considerably large with 3 μm at the point of maximum load for lubrication with SAE40, it decreases to about 2.5 μm and 2.0 μm for SAE30 and SAE20, respectively. For SAE10 the MOFT at the point of maximum load decreases further to about 1.5 μm; while from simulation still no asperity contact occurs for this case, in practical applications other effects not included here, like journal misalignment may lead to asperity contact.

The situation is quite different for a load of 70 MPa where all oils cannot avoid a certain amount of asperity contact and the MOFT consequently drops for all oils below 1.5 μm, however, for a different number of degrees crank angle. It is instructive to note that for a load of 70 MPa the MOFT changes not by the same amount between the different viscosity classes as for the 41 MPa load case. This can be explained by the choice of the elastic factor in the contact model that directly influences how much the asperity contact pressure increases for a given reduction in oil film thickness. Due to the use of a large elastic factor, already small changes in oil film thickness lead to large changes in the asperity contact pressure and, therefore, increase significantly the total load carrying capacity.

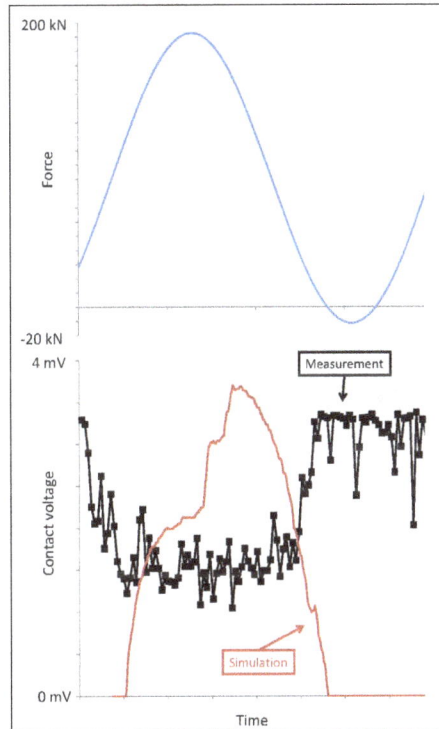

Plot of the measured contact voltage (black markers), whose decrease signals occuring metal-metal contact, in comparison to the metal-metal contact predicted from simulation (red line); as reference the blue curve depicts the applied load (top) for the bearing durability test.

In figure this increase in asperity contact pressure is depicted over shell angle at the shell edge for lubrication with the different oils and a load of 70 MPa. As can be seen, the asperity contact pressure increases from a maximum of about 2.5 MPa for the SAE40-oil to a maximum of 50 MPa for lubrication with the SAE10-oil and this explains the small MOFT changes that are seen for the different oils.

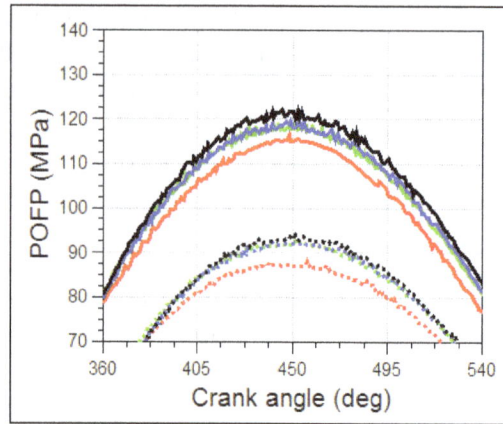

In the above figure, plot of the calculated peak oil film pressures (POFP) occurring in the test bearing during the 90 degrees crank cycle when the load is applied; dashed lines denote the results for an applied load of 41 MPa, solid lines the results for an applied load of 70 MPa; the red curves represents the results for SAE10, the green for SAE20, the blue for SAE30 and the black curves show the results for lubrication with SAE40-oil.

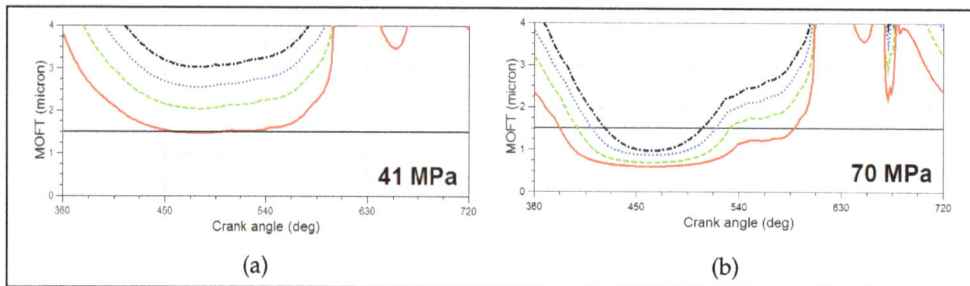

(a)   (b)

In the above figure, comparison of the minimum oil film thickness (denoted as MOFT, unit in μm) during a crank cycle in the test bearing for a load of (a) 41 MPa and (b) 70 MPa for 2000rpm journal speed calculated for the different oils: red solid line denotes SAE10, the green dashed line SAE20, the blue dotted line SAE30 and the black dash-dotted line SAE40. In addition the line at 1.5μm shows the threshold when asperity contact starts.

The presented asperity contact pressures occur only at the outermost bearing shell edges and drop sharply within a few millimetres in axial direction, as shown in figure exemplary for lubrication with SAE10 and a load of 70 MPa. The concentration of asperity contact in these areas is a consequence of the elastic deformation of the shell and journal under load and, therefore, these areas are particularly subject to wear. However, the exact shape of these areas depends also on the stiffness of the supporting structures.

This result further demonstrates the necessity of using realistic surface profiles in the simulation as otherwise asperity contact in these areas, and consequently also the friction moment, is largely overestimated. Performing an actual simulation for a load of 70 MPa and a lubrication with SAE10 without the discussed surface profile for the test bearing, simulation predicts a power loss due to asperity contact of 389 W (averaged over a full crank cycle), which exceeds by more than a factor six the value obtained for the simulation with the wear profile, namely 64 W, as shown in figure.

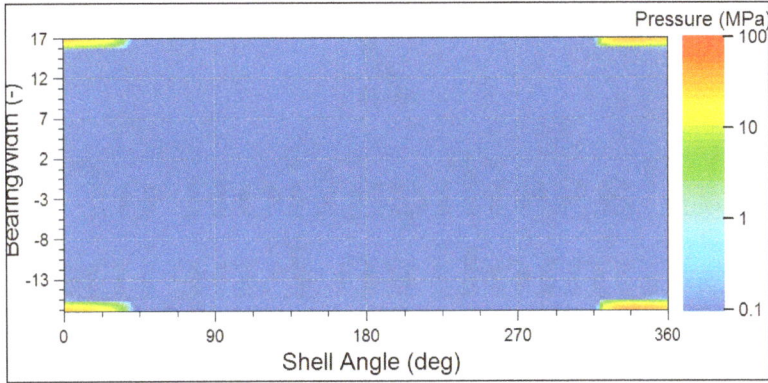

Map of the asperity contact pressures calculated for a load of 70 MPa and lubricated with SAE10-oil: the largest asperity contact pressures occur in very small areas at the bearing edges and drop sharply off within a few millimetres. The regions displayed in blue represent full film lubrication.

PACP[W]	PnoWPACP[W]	MSim[Nm]	M_{Sim}^{noWP}[Nm]	M_{LP06}[Nm]
64	389	4.7	6.9	(4.4-4.8)±0.5

Summary of the average friction moments predicted by simulation for a load of 70 MPa and lubrication with SAE10: the simulation using a surface profile for the test bearing is denoted as MSim and denoted as MnoWP Sim is the simulation without surface profile for the test bearing. The experimental values are denoted as MLP06 (range in brackets, ± measurement accuracy).

Consequently, simulation predicts for a load of 70 MPa and lubrication with SAE10 an average friction moment of 6.9 Nm which exceeds the experimental values by about 50%.

2
Classification of Automotive Engines

Automotive engines are broadly classified into internal combustion engines, steam engines and electric motors. Internal combustion engines are further divided into gasoline engines and diesel engines. The following chapter provides detailed information about these types of automotive engines.

Internal Combustion Engine

The internal combustion engine is an engine in which the burning of a fuel occurs in a confined space called a combustion chamber. This exothermic reaction of a fuel with an oxidizer creates gases of high temperature and pressure, which are permitted to expand. The defining feature of an internal combustion engine is that useful work is performed by the expanding hot gases acting directly to cause movement, for example by acting on pistons, rotors, or even by pressing on and moving the entire engine itself.

This contrasts with external combustion engines, such as steam engines, which use the combustion process to heat a separate working fluid, typically water or steam, which then in turn does work, for example by pressing on a steam actuated piston.

The term Internal Combustion Engine (ICE) is almost always used to refer specifically to reciprocating engines, Wankel engines and similar designs in which combustion is intermittent. However, continuous combustion engines, such as Jet engines, most rockets and many gas turbines are also internal combustion engines.

Internal combustion engines are seen mostly in transportation. Several other uses are for any portable situation where you need an non-electric motor. The largest application in this situation would be an Internal combustion engine driving an electric generator. That way, you can use standard electric tools driven by an internal combustion engine.

The advantages of these is the portability. It is more convenient using this type of engine in vehicles over electricity. Even in cases of hybrid vehicles, they still use an internal combustion engine to charge the battery. The disadvantage is the pollution that they put out. Not only the obvious, air pollution, but also pollution of broken or obsolete engines and waste parts, such as oil or rubber items that have to be discarded. Noise pollution is another factor, many internal combustion engines are very loud. Some are so loud, people need hearing protection to prevent damage their ears. Another disadvantage is size. It is very impractical to have small motors that can have any power. Electric motors are much more practical for this. That is why it is more likely to see an gas powered electric generator in an area that has no electricity to power smaller items.

Operation

All internal combustion engines depend on the exothermic chemical process of combustion: The reaction of a fuel, typically with air, although other oxidizers such as nitrous oxide may be employed.

The most common fuel in use today are made up of hydrocarbons and are derived from mostly petroleum. These include the fuels known as diesel fuel, gasoline, and petroleum gas, and rare use of propane gas. Most internal combustion engines designed for gasoline can run on natural gas or liquified petroleum gases without major modifications except for the fuel delivery components. Liquid and gaseous biofuels, such as Ethanol and biodiesel, a form of diesel fuel that is produced from crops that yield triglycerides such as soy bean oil, can also be used. Some can also run on Hydrogen gas.

All internal combustion engines must have a method for achieving ignition in their cylinders to create combustion. Engines use either a electrical method or a compression ignition system.

Gasoline Ignition Process

Electrical/Gasoline-type ignition systems generally rely on a combination of a lead-acid battery and an induction coil to provide a high voltage electrical spark to ignite the air-fuel mix in the engine's cylinders. This battery can be recharged during operation using an electricity-generating device, such as an alternator or generator driven by the engine. Gasoline engines take in a mixture of air and gasoline and compress to less than 170 psi and use a spark plug to ignite the mixture when it is compressed by the piston head in each cylinder.

Diesel Engine Ignition Process

Compression ignition systems, such as the diesel engine and HCCI (Homogeneous Charge Compression Ignition) engines, rely solely on heat and pressure created by the engine in its compression process for ignition. Compression that occurs is usually more than three times higher than a gasoline engine. Diesel engines will take in air only, and shortly before peak compression, a small quantity of diesel fuel is sprayed into the cylinder via a fuel injector that allows the fuel to instantly ignite. HCCI type engines will take in both air and fuel but will continue to rely on an unaided auto-combustion process due to higher pressures and heat. This is also why diesel and HCCI engines are also more susceptible to cold starting issues though they will run just as well in cold weather once started. Most diesels also have battery and charging systems however this system is secondary and is added by manufacturers as luxury for ease of starting, turning fuel on and off which can also be done via a switch or mechanical apparatus, and for running auxiliary electrical components and accessories. Most modern diesels, however, rely on electrical systems that also control the combustion process to increase efficiency and reduce emissions.

Energy

Once successfully ignited and burnt, the combustion products, hot gases, have more available energy than the original compressed fuel/air mixture (which had higher chemical energy). The available energy is manifested as high temperature and pressure which can be translated into work by the engine. In a reciprocating engine, the high pressure product gases inside the cylinders drive the engine's pistons.

Once the available energy has been removed, the remaining hot gases are vented (often by opening a valve or exposing the exhaust outlet) and this allows the piston to return to its previous position (Top Dead Center—TDC). The piston can then proceed to the next phase of its cycle, which varies between engines. Any heat not translated into work is normally considered a waste product, and is removed from the engine either by an air or liquid cooling system.

Parts

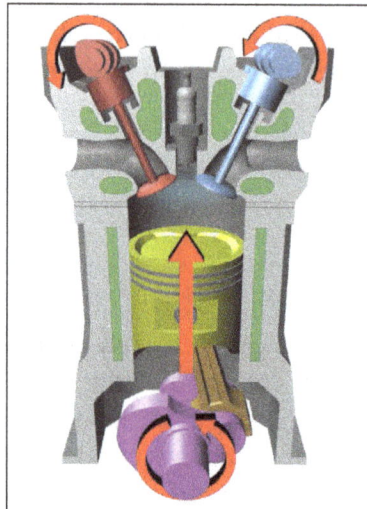

An illustration of several key components in a typical four-stroke engine.

The parts of an engine vary depending on the engine's type. For a four-stroke engine, key parts of the engine include the crankshaft (purple), one or more camshafts (red and blue) and valves. For a two-stroke engine, there may simply be an exhaust outlet and fuel inlet instead of a valve system. In both types of engines, there are one or more cylinders (gray and green) and for each cylinder there is a spark plug (darker-gray), a piston (yellow) and a crank (purple). A single sweep of the cylinder by the piston in an upward or downward motion is known as a stroke and the downward stroke that occurs directly after the air-fuel mix in the cylinder is ignited is known as a power stroke.

A Wankel engine has a triangular rotor that orbits in an epitrochoidal chamber around an eccentric shaft. The four phases of operation (intake, compression, power, exhaust) take place in separate locations, instead of one single location as in a reciprocating engine.

A Bourke Engine uses a pair of pistons integrated to a Scotch Yoke that transmits reciprocating force through a specially designed bearing assembly to turn a crank mechanism. Intake, compression, power, and exhaust all occur in each stroke of this yoke.

Classification

There is a wide range of internal combustion engines corresponding to their many varied applications. Likewise there is a wide range of ways to classify internal-combustion engines, some of which are listed below.

Although the terms sometimes cause confusion, there is no real difference between an "engine" and a "motor." At one time, the word "engine" meant any piece of machinery. A "motor" is any machine that produces mechanical power. Traditionally, electric motors are not referred to as "engines," but combustion engines are often referred to as "motors." (An electric engine refers to locomotive operated by electricity.)

With that said, one must understand that common usage does often dictate definitions. Many individuals consider engines as those things which generate their power from within, and motors as requiring an outside source of energy to perform their work. Evidently, the roots of the words seem to actually indicate a real difference. Further, as in many definitions, the root word only explains the beginnings of the word, rather than the current usage. It can certainly be argued that such is the case with the words motor and engine.

Principles of Operation

A 1906 gasoline engine.

Reciprocating

- Crude oil engine

- Two-stroke cycle

- Four-stroke cycle

- Hot bulb engine

- Poppet valves

- Sleeve valve
- Atkinson cycle
- Proposed
- Bourke engine
- Improvements
- Controlled Combustion Engine

Rotary

- Demonstrated:
  - Wankel engine
- Proposed:
  - Orbital engine
  - Quasiturbine
  - Rotary Atkinson cycle engine
  - Toroidal engine

Continuous Combustion

- Gas turbine
- Jet engine
- Rocket engine

Fuel and Oxidizer Types

Fuels used include petroleum spirit (North American term: Gasoline, British term: Petrol), autogas (liquified petroleum gas), compressed natural gas, hydrogen, diesel fuel, jet fuel, landfill gas, biodiesel, biobutanol, peanut oil and other vegoils, bioethanol, biomethanol (methyl or wood alcohol), and other biofuels. Even fluidized metal powders and explosives have seen some use. Engines that use gases for fuel are called gas engines and those that use liquid hydrocarbons are called oil engines. However, gasoline engines are unfortunately also often colloquially referred to as "gas engines."

The main limitations on fuels are that the fuel must be easily transportable through the fuel system to the combustion chamber, and that the fuel release sufficient energy in the form of heat upon combustion to make use of the engine practical.

The oxidizer is typically air, and has the advantage of not being stored within the vehicle, increasing the power-to-weight ratio. Air can, however, be compressed and carried aboard a vehicle. Some submarines are designed to carry pure oxygen or hydrogen peroxide to make them air-independent. Some race cars carry nitrous oxide as oxidizer. Other chemicals, such as chlorine or fluorine, have seen experimental use; but most are impractical.

Diesel engines are generally heavier, noisier, and more powerful at lower speeds than gasoline engines. They are also more fuel-efficient in most circumstances and are used in heavy road vehicles, some automobiles (increasingly more so for their increased fuel efficiency over gasoline engines), ships, railway locomotives, and light aircraft. Gasoline engines are used in most other road vehicles including most cars, motorcycles, and mopeds. Note that in Europe, sophisticated diesel-engined cars have become quite prevalent since the 1990s, representing around 40 percent of the market. Both gasoline and diesel engines produce significant emissions. There are also engines that run on hydrogen, methanol, ethanol, liquefied petroleum gas (LPG), and biodiesel. Paraffin and tractor vaporising oil (TVO) engines are no longer seen.

Hydrogen

Some have theorized that in the future hydrogen might replace such fuels. Furthermore, with the introduction of hydrogen fuel cell technology, the use of internal combustion engines may be phased out. The advantage of hydrogen is that its combustion produces only water. This is unlike the combustion of fossil fuels, which produce carbon dioxide, a principle cause of global warming, carbon monoxide resulting from incomplete combustion, and other local and atmospheric pollutants such as sulfur dioxide and nitrogen oxides that lead to urban respiratory problems, acid rain, and ozone gas problems. However, free hydrogen for fuel does not occur naturally, burning it liberates less energy than it takes to produce hydrogen in the first place by the simplest and most widespread method, electrolysis. Although there are multiple ways of producing free hydrogen, those require converting currently combustible molecules into hydrogen, so hydrogen does not solve any energy crisis, moreover, it only addresses the issue of portability and some pollution issues. The big disadvantage of hydrogen in many situations is its storage. Liquid hydrogen has extremely low density- 14 times lower than water and requires extensive insulation, whilst gaseous hydrogen requires very heavy tankage. Although hydrogen has a higher specific energy, the volumetric energetic storage is still roughly five times lower than petrol, even when liquified. (The "Hydrogen on Demand" process, designed by Steven Amendola, creates hydrogen as it is needed, but this has other issues, such as the raw materials being relatively expensive.) Other fuels that are kinder on the environment include biofuels. These can give no net carbon dioxide gains.

Cylinders

One-cylinder gasoline engine.

Internal combustion engines can contain any number of cylinders with numbers between one and twelve being common, though as many as 36 (Lycoming R-7755) have been used. Having more cylinders in an engine yields two potential benefits: First, the engine can have a larger displacement with smaller individual reciprocating masses (that is, the mass of each piston can be less) thus making a smoother running engine (since the engine tends to vibrate as a result of the pistons moving up and down). Second, with a greater displacement and more pistons, more fuel can be combusted and there can be more combustion events (that is, more power strokes) in a given period of time, meaning that such an engine can generate more torque than a similar engine with fewer cylinders. The down side to having more pistons is that, over all, the engine will tend to weigh more and tend to generate more internal friction as the greater number of pistons rub against the inside of their cylinders. This tends to decrease fuel efficiency and rob the engine of some of its power. For high performance gasoline engines using current materials and technology (such as the engines found in modern automobiles), there seems to be a break point around 10 or 12 cylinders, after which addition of cylinders becomes an overall detriment to performance and efficiency, although exceptions such as the W16 engine from Volkswagen exist.

- Most car engines have four to eight cylinders, with some high performance cars having ten, twelve, or even sixteen, and some very small cars and trucks having two or three. In previous years, some quite large cars, such as the DKW and Saab 92, had two cylinder, two stroke engines.

- Radial aircraft engines, now obsolete, had from three to 28 cylinders, such as the Pratt & Whitney R-4360. A row contains an odd number of cylinders, so an even number indicates a two or four-row engine. The largest of these was the Lycoming R-7755 with 36 cylinders (four rows of nine cylinders) but never entered production.

- Motorcycles commonly have from one to four cylinders, with a few high performance models having six (though some "novelties" exist with 8, 10, and 12).

- Snowmobiles usually have two cylinders. Some larger (not necessarily high-performance, but also touring machines) have four.

- Small portable appliances such as chainsaws, generators and domestic lawn mowers most commonly have one cylinder, although two-cylinder chainsaws exist.

Ignition System

Internal combustion engines can be classified by their ignition system. The point in the cycle at which the fuel/oxidizer mixture are ignited has a direct effect on the efficiency and output of the ICE. For a typical 4 stroke automobile engine, the burning mixture has to reach its maximum pressure when the crankshaft is 90 degrees after TDC (Top dead centre). The speed of the flame front is directly affected by compression ratio, fuel mixture temperature and octane or cetane rating of the fuel. Modern ignition systems are designed to ignite the mixture at the right time to ensure the flame front doesn't contact the descending piston crown. If the flame front contacts the piston, pinking or knocking results. Leaner mixtures and lower mixture pressures burn more slowly requiring more advanced ignition timing. Today most engines use an electrical or compression heating system for ignition. However outside flame and hot-tube systems have been used historically.

Nikola Tesla gained one of the first patents on the mechanical ignition system with U.S. Patent 609250, "Electrical Igniter for Gas Engines," on August 16, 1898.

Fuel Systems

Fuels burn faster, and more completely when they have lots of surface area in contact with oxygen. In order for an engine to work efficiently the fuel must be vaporized into the incoming air in what is commonly referred to as a fuel air mixture. There are two commonly used methods of vaporizing fuel into the air, one is the carburetor and the other is fuel injection.

Often for simpler, reciprocating engines a carburetor is used to supply fuel into the cylinder. However, exact control of the correct amount of fuel supplied to the engine is impossible. Carburetors are the current most widespread fuel mixing device used in lawnmowers and other small engine applications. Prior to the mid-1980s, carburetors were also common in automobiles.

Larger gasoline engines such as those used in automobiles have mostly moved to fuel injection systems. Diesel engines always use fuel injection.

Autogas (LPG) engines use either fuel injection systems or open or closed loop carburetors.

Other internal combustion engines like jet engines use burners, and rocket engines use various different ideas including impinging jets, gas/liquid shear, preburners, and many other ideas.

Engine Configuration

Internal combustion engines can be classified by their configuration which affects their physical size and smoothness (with smoother engines producing less vibration). Common configurations include the straight or inline configuration, the more compact V configuration and the wider but smoother flat or boxer configuration. Aircraft engines can also adopt a radial configuration which allows more effective cooling. More unusual configurations, such as "H," "U," "X," or "W" have also been used.

Multiple-crankshaft configurations do not necessarily need a cylinder head at all, but can instead have a piston at each end of the cylinder, called an opposed piston design. This design was used in the Junkers Jumo 205 diesel aircraft engine, using two crankshafts, one at either end of a single bank of cylinders, and most remarkably in the Napier Deltic diesel engines, which used three crankshafts to serve three banks of double-ended cylinders arranged in an equilateral triangle with the crankshafts at the corners. It was also used in single-bank locomotive engines, and continues to be used for marine engines, both for propulsion and for auxiliary generators. The Gnome Rotary engine, used in several early aircraft, had a stationary crankshaft and a bank of radially arranged cylinders rotating around it.

Engine Capacity

An engine's capacity is the displacement or swept volume by the pistons of the engine. It is generally measured in liters (L) or cubic inches (c.i. or in³) for larger engines and cubic centimeters (abbreviated to cc) for smaller engines. Engines with greater capacities are usually more powerful and provide greater torque at lower rpm but also consume more fuel.

Apart from designing an engine with more cylinders, there are two ways to increase an engine's capacity. The first is to lengthen the stroke and the second is to increase the piston's diameter. In either case, it may be necessary to make further adjustments to the fuel intake of the engine to ensure optimal performance.

An engine's quoted capacity can be more a matter of marketing than of engineering. The Morris Minor 1000, the Morris 1100, and the Austin-Healey Sprite Mark II were all fitted with a BMC A-Series engine of the same stroke and bore according to their specifications, and were from the same maker. However the engine capacities were quoted as 1000cc, 1100cc, and 1098cc respectively in the sales literature and on the vehicle badges.

Lubrication Systems

There are several different types of lubrication systems used. Simple two-stroke engines are lubricated by oil mixed into the fuel or injected into the induction stream as a spray. Early slow speed stationary and marine engines were lubricated by gravity from small chambers, similar to those used on steam engines at the time, with an engine tender refilling these as needed. As engines were adapted for automotive and aircraft use, the need for a high power to weight ratio lead to increased speeds, higher temperatures, and greater pressure on bearings, which, in turn, required pressure lubrication for crank bearing and connecting rod journals, provided either by a direct lubrication from a pump, or indirectly by a jet of oil directed at pickup cups on the connecting rod ends, which had the advantage of providing higher pressures as engine speed increased.

Engine Pollution

Generally internal combustion engines, particularly reciprocating internal combustion engines, produce moderately high pollution levels, due to incomplete combustion of carbonaceous fuel, leading to carbon monoxide and some soot along with oxides of nitrogen and sulfur and some unburnt hydrocarbons depending on the operating conditions and the fuel/air ratio. The primary causes of this are the need to operate near the stoichiometric ratio for petrol engines in order to achieve combustion (the fuel would burn more completely in excess air) and the "quench" of the flame by the relatively cool cylinder walls.

Diesel engines produce a wide range of pollutants including aerosols of many small particles (PM10) that are believed to penetrate deeply into human lungs. Engines running on liquefied petroleum gas (LPG) are very low in emissions as LPG burns very clean and does not contain sulphur or lead.

- Many fuels contain sulfur leading to sulfur oxides (SOx) in the exhaust, promoting acid rain.

- The high temperature of combustion creates greater proportions of nitrogen oxides (NOx), demonstrated to be hazardous to both plant and animal health.

- Net carbon dioxide production is not a necessary feature of engines, but since most engines are run from fossil fuels this usually occurs. If engines are run from biomass, then no net carbon dioxide is produced as the growing plants absorb as much, or more carbon dioxide while growing.

- Hydrogen engines need only produce water, but when air is used as the oxidizer nitrogen oxides are also produced.

Internal Combustion Engine Efficiency

The efficiency of various types of internal combustion engines vary. It is generally accepted that most gasoline fueled internal combustion engines, even when aided with turbochargers and stock efficiency aids, have a mechanical efficiency of about 20 percent. Most internal combustion engines waste about 36 percent of the energy in gasoline as heat lost to the cooling system and another 38 percent through the exhaust. The rest, about six percent, is lost to friction. Most engineers have not been able to successfully harness wasted energy for any meaningful purpose, although there are various add on devices and systems that can greatly improve combustion efficiency.

Hydrogen Fuel Injection, or HFI, is an engine add on system that is known to improve fuel economy of internal combustion engines by injecting hydrogen as a combustion enhancement into the intake manifold. Fuel economy gains of 15 percent to 50 percent can be seen. A small amount of hydrogen added to the intake air-fuel charge increases the octane rating of the combined fuel charge and enhances the flame velocity, thus permitting the engine to operate with more advanced ignition timing, a higher compression ratio, and a leaner air-to-fuel mixture than otherwise possible. The result is lower pollution with more power and increased efficiency. Some HFI systems use an on board electrolyzer to generate the hydrogen used. A small tank of pressurized hydrogen can also be used, but this method necessitates refilling.

There has also been discussion of new types of internal combustion engines, such as the Scuderi Split Cycle Engine, that utilize high compression pressures in excess of 2000 psi and combust after top-dead-center (the highest & most compressed point in a internal combustion piston stroke). Such engines are expected to achieve efficiency as high as 50-55%.

Gasoline Engine

Gasoline engine is any of a class of internal-combustion engines that generate power by burning a volatile liquid fuel (gasoline or a gasoline mixture such as ethanol) with ignition initiated by an electric spark. Gasoline engines can be built to meet the requirements of practically any conceivable power-plant application, the most important being passenger automobiles, small trucks and buses, general aviation aircraft, outboard and small inboard marine units, moderate-sized stationary pumping, lighting plants, machine tools, and power tools. Four-stroke gasoline engines power the vast majority of automobiles, light trucks, medium-to-large motorcycles, and lawn mowers. Two-stroke gasoline engines are less common, but they are used for small outboard marine engines and in many handheld landscaping tools such as chain saws, hedge trimmers, and leaf blowers.

Engine Types

Gasoline engines can be grouped into a number of types depending on several criteria, including their application, method of fuel management, ignition, piston-and-cylinder or rotor arrangement, strokes per cycle, cooling system, and valve type and location. In a piston-and-cylinder engine the pressure produced by combustion of gasoline creates a force on the head of a piston that moves the length of the cylinder in a reciprocating, or back-and-forth, motion. This force drives the piston away from the head of the cylinder and performs work. The rotary engine, also called the Wankel engine, does not have conventional cylinders fitted with reciprocating pistons. Instead, the gas pressure acts on the surfaces of a rotor, causing the rotor to turn and thus perform work.

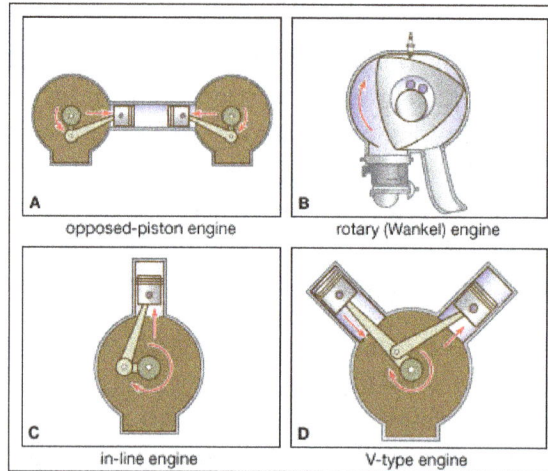

Four Types of Gasoline Engines.

Piston-and-Cylinder Engines

Most gasoline engines are of the reciprocating piston-and-cylinder type. The essential components of the piston-and-cylinder engine are shown in the figure. Almost all engines of this type follow either the four-stroke cycle or the two-stroke cycle.

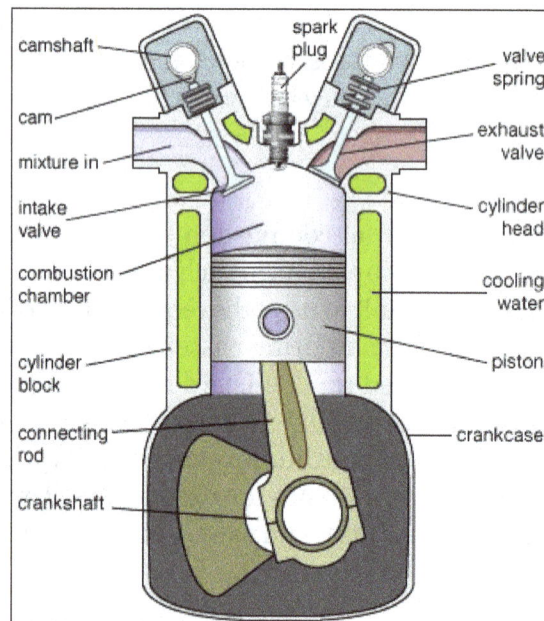

Typical piston-cylinder arrangement of a gasoline engine.

Four-stroke Cycle

Of the different techniques for recovering the power from the combustion process, the most important so far has been the four-stroke cycle, a conception first developed in the late 19th century.

The four-stroke cycle is illustrated in the figure. With the inlet valve open, the piston first descends on the intake stroke. An ignitable mixture of gasoline vapour and air is drawn into the cylinder by the partial vacuum thus created. The mixture is compressed as the piston ascends on the compression stroke with both valves closed. As the end of the stroke is approached, the charge is ignited by an electric spark. The power stroke follows, with both valves still closed and the gas pressure, due to the expansion of the burned gas, pressing on the piston head or crown. During the exhaust stroke the ascending piston forces the spent products of combustion through the open exhaust valve. The cycle then repeats itself. Each cycle thus requires four strokes of the piston—intake, compression, power, and exhaust—and two revolutions of the crankshaft.

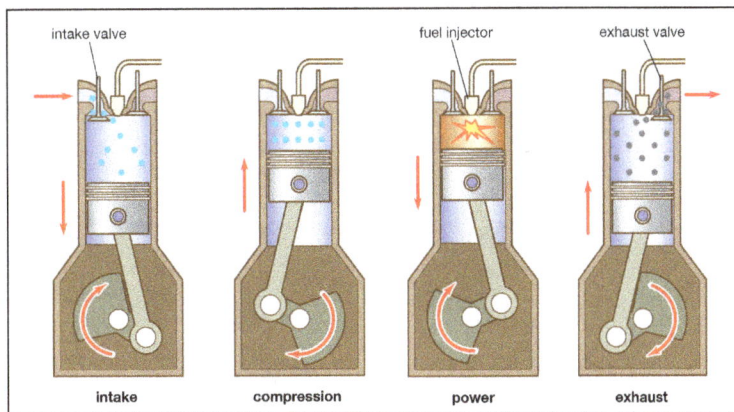

Internal-combustion engine: four-stroke cycle.

A disadvantage of the four-stroke cycle is that only half as many power strokes are completed as in the two-stroke cycle and only half as much power can be expected from an engine of a given size at a given operating speed. The four-stroke cycle, however, provides more positive clearing out of exhaust gases (scavenging) and reloading of the cylinders, reducing the loss of fresh charge to the exhaust.

Two-stroke Cycle

In the original two-stroke cycle (as developed in 1878), the compression and power stroke of the four-stroke cycle are carried out without the inlet and exhaust strokes, thus requiring only one revolution of the crankshaft to complete the cycle. The fresh fuel mixture is forced into the cylinder through circumferential ports by a rotary blower in the two-stroke-cycle engine of a so-called uniflow type. The exhaust gases pass through poppet valves in the cylinder head that are opened and closed by a cam-follower mechanism. The valves are timed to begin opening toward the end of the power stroke, after the cylinder pressure has dropped appreciably. The inlet ports in the cylinder wall start to uncover after the exhaust opening has decreased the cylinder pressure to the inlet pressure produced by the blower. The exhaust valves are allowed to remain open for a few degrees of crank rotation after the inlet ports have been covered by the rising piston on the compression stroke, thus allowing the persistency of flow to scavenge the cylinder more thoroughly. The compression and power strokes are similar to those of the four-stroke engine.

A simplified version of the two-stroke-cycle engine was developed some years later (introduced in 1891) by using crankcase compression to pump the fresh charge into the cylinder. Instead of intake ports extending entirely around the lower cylinder wall, this engine has intake ports only halfway around; a second set of ports starts a little higher in the cylinder wall in the other half of the cylinder bore. These larger ports lead to the exhaust system. The inlet ports connect to a transfer passage leading to the fully enclosed crankcase. A spring-loaded inlet valve admits air into the crankcase on the upward, or compression, stroke of the piston. Air trapped in the crankcase is compressed by the descent of the piston on its power stroke. The piston thus uncovers the exhaust ports near the end of the power stroke, and slightly later it uncovers the inlet, or transfer, port on the opposite side of the cylinder to admit the compressed fresh mixture from the crankcase. The top face of the piston is designed to provide a deflector or baffle that directs the fresh load upward on the inlet side of the cylinder and then downward on the exhaust side, thus pushing the spent gases of the previous cycle out through the exhaust port on that side. This outflow continues after the inlet ports are covered by the rising piston on the compression stroke, until the exhaust ports are covered and compression of the fresh load begins. This loading process, called loop scavenging, is the simplest known method of replacing the exhaust products with a fresh mixture and creating a cycle with only compression and power strokes.

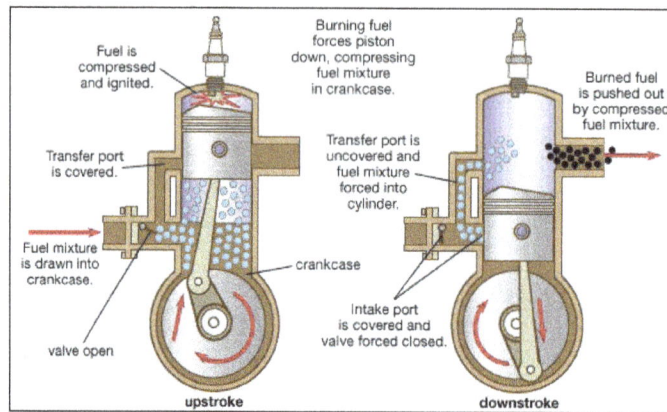

Blower-scavenged, two-stroke-cycle engine with uniflow scavenging.

Such a system is used in many small gasoline engines (e.g., small outboard motors) and for gasoline-powered appliances. A disadvantage is that the return flow of the gases causes a slight loss of fresh charge through the exhaust ports. Because of this loss, carburetor engines operating on the two-stroke cycle lack the fuel economy of four-stroke engines. The loss can be avoided by equipping them with fuel-injection systems instead of carburetors and injecting the fuel directly into the cylinders after scavenging. Such an arrangement is attractive as a means of attaining high power output from a relatively small engine, and development of the turbocharger for this application holds promise of further improvement.

Opposed-piston Engines

The opposed-piston engine also provides uniflow scavenging. This engine has two pistons moving in opposite directions in the same cylinder. Two sets of ports extending entirely around the cylinder bore are located so that one set is covered and uncovered by one piston and the other set is controlled by the second piston. A second crankshaft, to which the upper pistons are attached, is located at the top of the engine, and the two shafts are connected by gears.

The opposed-piston design has two major advantages: reciprocating masses move in opposite directions, providing excellent balance; and the poppet valves necessary in other uniflow-scavenged two-stroke-cycle engines are eliminated.

Rotary (Wankel) Engines

The rotary-piston internal-combustion engine developed in Germany is radically different in structure from conventional reciprocating piston engines. This engine was conceived by Felix Wankel, a specialist in the design of sealing devices, and experimental units were built and tested by a German firm beginning in 1956. Instead of pistons that move up and down in cylinders, the Wankel engine has an equilateral triangular orbiting rotor. The rotor turns in a closed chamber, and the three apexes of the rotor maintain a continuous sliding contact with the curved inner surface of the casing. The curve-sided rotor forms three crescent-shaped chambers between its sides and the curved wall of the casing. The volumes of the chambers vary with rotor position. Maximum volume is attained in each chamber when the side of the rotor forming it is parallel with the minor diameter of the casing; the volume is reduced to a minimum when the rotor side is parallel with the major diameter. Shallow pockets recessed in the flank of the rotor control the shape of the combustion chambers and establish the compression ratio of the engine.

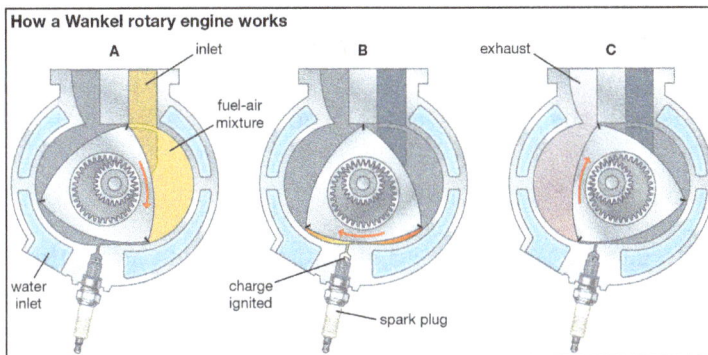

Wankel rotary engine One cycle of the Wankel rotary engine, showing
(A) intake, (B) ignition, and (C) exhaust stages.

In turning about its central axis, the rotor must follow a circular orbit about the geometric centre of the casing. The necessary orbiting rotation is attained by means of a central bore in the rotor in which an internal gear is fitted to mesh with a stationary pinion fixed immovably to the centre of the casing. The rotor is guided by fitting its central bore to an eccentric formed on the output shaft that passes through the centre of the stationary pinion. This eccentric also harnesses the rotor to the shaft so that torque is applied when gas pressure is exerted against the rotor flanks as the fuel and air charges burn. A 3-to-1 gear ratio causes the output shaft to turn three times as fast as the rotor turns about the eccentric. Each quarter turn of the rotor completes an expansion or a compression, permitting intake, compression, expansion, and exhaust to be accomplished during one turn of the rotor. The only moving parts are the rotor and the output shaft.

The fuel mixture is supplied by a carburetor and enters the combustion chambers through an intake port in one of the end plates of the casing. An exhaust port is formed in one of the flattened sides of the casing wall, and a spark plug is located in a pocket communicating with the chambers through a small throat in the opposite side of the casing wall.

The rotor and its gears and bearings are lubricated and cooled by oil circulating through the hollow rotor. The apex vanes are lubricated by a small amount of oil added to the fuel in proportions as low as 1 to 200. Water is circulated through cooling jackets in the casing, the entrance to which is located adjacent to the spark plug, where the temperature tends to be highest.

Maintaining pressure-tight joints by suitable seals at the apexes and on the end faces of the rotor is a major design problem. Radial sliding vanes are fitted in slots at the three apex edges and kept in contact with the casing by expander springs. The end faces of the rotor are sealed by arc-shaped segmental rings fitted in grooves close to the curved edges of the rotor and pressed against the casing by flat springs.

The major advantages of the Wankel engine are its small space requirements and low weight per horsepower, smooth and vibrationless operation, quiet operation, and low manufacturing costs resulting from mechanical simplicity. The absence of inertial forces from reciprocating parts and the elimination of spring-closed poppet valves permit operation at much higher speed than is practical for reciprocating piston engines, an advantage because shaft speed must be high for optimum performance. The induction of fresh fuel mixture and exhaust are more effective because the ports are opened and closed more rapidly than with poppet valves, and gas flow through them is almost continuous. Heat transfer and the resulting cooling requirement are low because the jacketed surface is small. Lower weight and a lower centre of gravity make it much safer in an automobile in the event of a collision. However, competitive fuel economies and the higher development and manufacturing costs of meeting emission standards have limited the use of the Wankel engine in production vehicles, with only the Mazda Motor Corporation marketing any substantial number.

Engine Construction and Operation

The overall structure of a gasoline engine depends almost entirely upon the intended application. Apart from the type of cycle (two- or four-stroke), the provision for mounting is the main structural difference among automotive, marine, stationary, and aviation engines. When a clutch and transmission are used, as in automobiles, the engine is commonly of the so-called unit-power-plant type with a bell-shaped housing surrounding the flywheel and attached to the rear flange of the cylinder block integral with, or attached to, the transmission gear case. The clutch is incorporated in the flywheel of the engine. Three-point suspension is used in such engines; that is to say, projections on each side of the bell housing fit into the vehicle side-frame members, and a central tubular extension at the centre of the front end of the cylinder block attaches to the front cross member of the frame. This construction permits some flexing of the vehicle frame without stressing the basic structure of the engine.

The following description of general engine construction indicates the essential components of a piston-and-cylinder engine and introduces the nomenclature of the various parts. The four-stroke-cycle automobile engine is used as the basic type.

Cylinder Block

The main structural member of all automotive engines is a cylinder block that usually extends upward from the centre line of the main support for the crankshaft to the junction with the cylinder head. The block serves as the structural framework of the engine and carries the mounting pad by which the engine is supported in the chassis. Large, stationary power-plant engines and marine engines are built up from a foundation, or bedplate, and have upper and lower crankcases that are separate from the cylinder assemblies. The cylinder block of an automobile engine is a casting with appropriate machined surfaces and threaded holes for attaching the cylinder head, main bearings, oil pan, and other units. The crankcase is formed by the portion of the cylinder block below the cylinder bores and the stamped or cast metal oil pan that forms the lower enclosure of the engine and also serves as a lubricating oil reservoir, or sump.

The cylinders are openings of circular cross section that extend through the upper portion of the block, with interior walls bored and polished to form smooth, accurate bearing surfaces. The cylinders of heavy-duty engines are usually fitted with removable liners made of metal that is more wear-resistant than that used in the block casting.

There are two arrangements of cylinders in common automotive use—the vertical, or in-line, type and the V type. The in-line engine has a single row of cylinders extending vertically upward from the crankcase and aligned with the crankshaft main bearings. The V type has two rows of cylinders, usually forming an angle of 60° or 90° between the two banks. V-8 engines (eight cylinders) are usually of the 90° type. Some small six-cylinder aviation engines have horizontally opposed cylinders.

A passage bored lengthwise in the block houses the camshaft that operates the valves. The location of camshafts for most automotive applications is overhead—overhead cam (OHC) or dual overhead cam (DOHC). A gear, chain, or belt compartment for the camshaft drive from the crankshaft is formed between the front or rear end of the block and a cover plate. On virtually all modern engines, a toothed belt is used to ensure accurate and responsive control of the valve train. The bell housing is formed at the rear of the cylinder block to enclose the flywheel and provide for attachment of a transmission housing. Water jackets are formed around the cylinders with suitable cored connecting passages for circulation of the coolant.

The design of the cylinder block is affected by the location of the valves of the four-stroke-cycle engine and by the provision of cylinder ports in the two-stroke type. An overhead-valve engine, which has largely replaced the L-head type, has its valves entirely in the cylinder head. The cylinder block of the L-head engine is extended to one side of the cylinder bores, with the valve seats and passages for inlet and exhaust, together with the valve guides, formed in this extension of the block. The cylinder head then becomes merely a water-jacketed cover, providing threaded locations for the spark plugs and with its underside so profiled that a combustion chamber of desired size and shape is formed above each cylinder bore. The shape of the space forming the combustion chamber when the piston is at its closest approach to the cylinder head and the volume contained therein in relation to the piston displacement volume are extremely important in their effect on performance. The cylinder head of the valve-in-head engine is narrower and deeper and carries the valve seats, valve guides, and valve ports.

Combustion Chamber

The combustion chamber is defined by the size, location, and position of the piston within the cylinder. Bore is the inner diameter of the cylinder. The volume at bottom dead centre (VBDC) is defined as the volume occupied between the cylinder head and the piston face when the piston is farthest from the cylinder head. The volume at top dead centre (VTDC) is the volume occupied when the piston is closest to the cylinder head; the distance between the piston face and cylinder head at VTDC is called the clearance. The distance traveled by the piston between its VTDC and VBDC locations is the stroke. The ratio of VTDC to VBDC normalized to the VTDC value—i.e., (VBDC/VTDC):1—is the compression ratio of a reciprocating engine. Compression ratio is the most important factor affecting the theoretical efficiency of the engine cycle. Because increasing the compression ratio is the best way to improve efficiency, compression ratios on automobile engines have tended to increase. This requires stronger, more-durable materials. In practice, fuel ignition characteristics, often represented by octane number, limit engine compression ratios.

Pistons

The pistons are cup-shaped cylindrical castings of steel or aluminum alloy. The upper, closed end, called the crown, forms the lower surface of the combustion chamber and receives the force applied by the combustion gases. The outer surface is machined to fit the cylinder bore closely and is grooved to receive piston rings that seal the gap between the piston and the cylinder wall. In the upper piston grooves there are plain compression rings that prevent the combustion gases from blowing past the piston. The lower rings are vented to distribute and limit the amount of lubricant on the cylinder wall. Piston pin supports (bosses) are cast in opposite sides of the piston and hardened steel pins fitted into these bosses pass through the upper end of the connecting rod.

Connecting Rod and Crankshaft

A forged-steel connecting rod connects the piston to a throw (offset portion) of the crankshaft and converts the reciprocating motion of the piston to the rotating motion of the crank. The lower, larger end of the rod is bored to take a precision bearing insert lined with babbitt or other bearing metal and closely fitted to the crankpin. V-type engines usually have opposite cylinders staggered sufficiently to permit the two connecting rods that operate on each crank throw to be side by side. Some larger engines employ fork-and-blade rods with the rods in the same plane and cylinders exactly opposite each other.

Each connecting rod in an in-line engine or each pair of rods in a V-type engine is attached to a throw of the crankshaft. Each throw consists of a crankpin with a bearing surface, on which the connecting-rod bearing insert is fitted, and two radial cheeks that connect it to the portions of the crankshaft that turn in the main bearings, supported by the cylinder block. Sufficient throws are provided to serve all the cylinders, and the angles between them equal the angular firing intervals between the cylinders. The throws of a six-cylinder, four-stroke-cycle crankshaft are spaced 120° apart so that the six cylinders fire at equal intervals in two full rotations of the shaft. Those of an eight-cylinder engine are 90° apart. The position of each throw along the shaft depends upon the firing order of the cylinders. Firing sequence is chosen to distribute the power impulses along the length of the engine to minimize vibration. Consideration is also given to the fluid flow pattern in the intake and exhaust manifolds. The standard firing order for a six-cylinder engine is 1-5-3-6-2-4, which illustrates the practice of alternating successive impulses between the front and rear valves of the engine whenever possible. Balance is further improved by adding counterweights to the crankshaft to offset the eccentric masses of metal in the crank throws.

The crankshaft design also establishes the length of the piston stroke because the radial offset of each throw is equal to half the stroke imparted to the piston. The ratio of the piston stroke to the cylinder bore diameter is an important design consideration. In the early years of engine development, no logical basis for the establishment of this ratio existed, and a range from unity to $1^1/_2$ was used by different manufacturers. As engine speeds increased, however, and it became apparent that friction horsepower increased with piston speed rather than with crankshaft rotating speed, there began a trend toward short-stroke engines. Strokes were shortened to as much as 20 percent less than the bores.

From the requirement for the two-cylinder engine, a general rule for the layout of the throws of four-stroke-cycle multicylinder crankshafts can be expressed. Regardless of the number of cylinders, two pistons must arrive at top dead centre in unison so that a second cylinder is ready to fire exactly 360° after each cylinder fires. Half the cylinders will then fire during each turn of the crankshaft. To follow this rule, there must be an even number of cylinders in order that there may be pairs of cylinders whose pistons move in unison.

An eight-cylinder engine fires each time its crankshaft makes a quarter turn if the intervals between impulses are equal. The crankshaft for an eight-cylinder, in-line engine is designed with each of its eight throws a quarter turn away from another throw.

For best lengthwise balance, the cylinders whose pistons are in phase are the first and last cylinders of an in-line engine, the second and next to the last, continuing in that order with crank throws that are in alignment equidistant from the centre of the engine.

Valves, Pushrods and Rocker Arms

Valves for controlling intake and exhaust may be located overhead, on one side, on one side and overhead, or on opposite sides of the cylinder. These are all the so-called poppet, or mushroom, valves, consisting of a stem with one end enlarged to form a head that permits flow through a passage surrounding the stem when raised from its seat and that prevents flow when the head is moved down to contact the valve seat formed in the cylinder block. Another group of engines uses sliding valves that are usually of the sleeve type surrounding the cylinder bore.

The valve-in-head engine has pushrods that extend upward from the cam followers to rocker arms mounted on the cylinder head that contact the valve stems and transmit the motion produced by the cam profile to the valves. Clearance (usually termed tappet clearance) must be maintained between the ends of the valve stems and the lifter mechanism to assure proper closing of the valves when the engine temperature changes. This is done by providing pushrod length adjustment or by the use of hydraulic lifters.

Noisy and erratic valve operation can be eliminated with entirely mechanical valve-lifter linkage only if the tappet clearance between the rocker arms and the valve stems is closely maintained at the specified value for the engine as measured with a thickness gauge. Hydraulic valve lifters, now commonly used on automobile engines, eliminate the need for periodic adjustment of clearance.

The hydraulic lifter comprises a cam follower that is moved up and down by contact with the cam profile, and an inner bore into which the valve lifter is closely fitted and retained by a spring clip. The valve lifter, in turn, is a cup closed at the top by a freely moving cylindrical plug that has a socket at the top to fit the lower end of the pushrod. This plug is pushed upward by a light spring that is merely capable of taking up the clearance between the valve stem and the rocker arm. A small hole is drilled in the bottom of the valve-lifter cup to admit lubricating oil that enters the cam follower from the engine lubricating system through a passage in the cylinder block. A small steel ball serves as a check valve to admit the oil into the valve-lifter cup but prevent its escape. When the clearance in the entire linkage between the cam profile and the valve stem is being taken up by the spring in the valve lifter, oil flows into the lifter chamber, past the ball check, and is trapped there to maintain this no-clearance condition as the engine operates. Expansion or contraction of the valve linkage is compensated by oil seepage from the lifter to correct for expansion of parts and oil flow into the chamber if clearance tends to be produced between the pushrod and the lifter. Complete closure of the valve is then assured at all times without tappet noise.

The intake valve must be open while the piston is descending on the intake stroke of the piston, and the exhaust valve must be open while the piston is rising on the exhaust stroke. It would seem, therefore, that the opening and closing of the two valves would occur at the appropriate top and bottom dead-centre points of the crankshaft. The time required for the valves to open and close, however, and the effects of high speed on the starting and stopping of the flow of the gases require that for optimum performance the opening events occur before the crankshaft dead-centre positions and that the closing events be delayed until after dead centre.

All four valve events—inlet opening, inlet closing, exhaust opening, and exhaust closing—are accordingly displaced appreciably from the top and bottom dead centres. Opening events are earlier and closing events are later to permit ramps to be incorporated in the cam profiles to allow gradual initial opening and final closing to avoid slamming of the valves. Ramps are provided to start the lift gradually and to slow down the valve before it contacts its seat. Early opening and late closure are also for the purpose of using the inertia or persistence of flow of the gases to assist in filling and emptying the cylinder.

Camshaft

The camshaft, which opens and closes the valves, is driven from the crankshaft by a chain drive or gears on the front end of the engine. Because one turn of the camshaft completes the valve operation for an entire cycle of the engine and the four-stroke-cycle engine makes two crankshaft revolutions to complete one cycle, the camshaft turns half as fast as the crankshaft. It is located above and to one side of the crankshaft, which places it directly under the valves of the L-head engine or the pushrods that extend down from the rocker arms of the valve-in-head engine. Because of the long pushrods and the rocker arms, the speed of the valve-in-head engine is limited to that at which the cam followers can remain in contact with the cams when the valves are closing. Above that limiting speed the valves are said to float, and their motion tends to become erratic. For this reason, the overhead-camshaft engine is quite popular. Located immediately above the valves, this type of camshaft is driven either by a vertical shaft and bevel gears or by a cog belt.

Flywheel

The cycle of the internal-combustion engine is such that torque (turning force) is applied only intermittently as each cylinder fires. Between these power impulses, the pistons rising on compression and the opposition to rotation caused by the load carried by the engine apply negative torque. The alternating acceleration caused by the power impulse and deceleration caused by compression result in nonuniform rotation. To counter this tendency to slow down and speed up is the function of the flywheel, attached to one end of the crankshaft. The flywheel consists of a heavy circular cast-iron disk with a hub for attachment to the engine. Its heavy rotating mass has sufficient momentum to oppose all changes in its rotational speed and to force the crankshaft to turn steadily at this speed. The engine thus runs smoothly with no evidence of rotational pulsations. The outer rim of the flywheel usually carries gear teeth so as to mesh with the starter motor. The driving component of a clutch or fluid coupling for the transmission may be incorporated in the flywheel.

Bearings

The crankshaft has bearing surfaces on each crank throw and three or more main bearings. These are heavily loaded because of the reciprocating forces at each cylinder applied to the crankshaft and the weight of the crankshaft and flywheel. All but the smallest engines use split-shell bearings, usually made of bronze with babbitt metal linings. The surface material is sufficiently soft to minimize the possibility of scoring the crankshaft in the event of inadequate lubrication. The smallest engines usually have cast-babbitt bearings. A small amount of bearing clearance is necessary to permit an oil film to separate the surfaces.

Ignition

Ignition Systems

Electric ignition systems may be classified as magneto, battery-and-coil, and solid-state ignition systems. Although these are similar in basic principle, the magneto is self-contained and requires only the spark plugs and connecting wires to complete the system, whereas the battery-and-coil and solid-state ignition systems involve several separate components.

A magneto is a fixed-magnet, alternating-current generator designed to produce sufficient voltage to fire the spark plugs. A high-tension magneto is entirely self-contained and requires only spark plugs, wires, and switches to meet ignition requirements.

The battery-and-coil system consists of a battery, one terminal of which is grounded while the other leads through a switch to the primary winding of the coil, and then to a circuit breaker where it is again grounded. Rotation of the circuit-breaker cam opens and closes the primary circuit. The secondary circuit, consisting of several thousand turns of fine wire, leads to the rotor of the distributor, which acts as a rotary switch, selecting the spark plug to be placed in the circuit. Each plug is connected to one of the outer terminals of the distributor to receive an electrical impulse in proper sequence. When the primary circuit is broken, a high potential (up to 20,000 volts) is developed in the secondary winding and conducted to the appropriate spark plug.

The high voltage for the spark plug may also be produced by a capacitor discharge ignition system. Such a system consists of a source of 250 to 300 volts direct-current power applied to a storage capacitor, a device for storing an electric charge. A lead from the capacitor goes to one side of the spark coil primary through cam-actuated breaker points or an electronic switching device. At the instant this switching device establishes a contact, the capacitor discharges through the primary of the spark coil, and an instantaneous high voltage is delivered to the distributor and thence to the spark plug. The capacitor discharge system provides a more intense spark, thus improving the start-up of a cold or flooded engine. It continues to fire the plugs when they are fouled by carbon or other deposits or when the spark gap has widened because of erosion of the points. Other notable advantages include increased spark plug life, improved firing over a wider speed range, and better moisture tolerance.

Solid-state ignition systems, unlike battery-and-coil systems that use a distributor, use an electronic module to collect information from engine sensors, compute engine operating parameters, and control ignition discharge to a separate coil for each spark plug. The electronic control module activates a transistor to break the ground circuit leading to each plug's coil, thereby causing a spark. In addition to eliminating the high-voltage spark plug wires, electronics allow for more precise control of ignition timing, which improves fuel efficiency, reduces emissions, and increases power.

Spark Plugs

The spark plug is an important component of the ignition system and is the one that must operate under the most severe conditions. Because it is exposed to combustion chamber temperatures and pressures and contaminating products of combustion, it requires more service attention and is usually the shortest-lived component of the gasoline engine. It consists of a steel shell threaded to fit a standard 14-mm hole in the cylinder head. Spark plugs may use a gasket or a tapered seat to ensure a gastight fit between cylinder head and plug. A fused ceramic insulating element is molded into the plug body, and the steel centre electrode passes through the insulator up to the connector to which the high-voltage lead from the distributor is attached. The other electrode is welded to the metal body of the plug, which is grounded to the cylinder head. Electrodes are found in a number of configurations and are made of a variety of alloys.

In application it is essential that the spark gap be as specified for the particular engine. Gauges are available to aid in making this adjustment by bending the ground electrode as required. Manufacturers specify gaps ranging from 0.508 to 1.016 mm between the centre electrode and the ground electrode. If the plug gap is too large, the possibility of misfiring increases. If the gap is too small, the spark will not be sufficiently intense. Gap growth from erosion of the electrodes may be corrected. Modern spark plugs often incorporate a resistor to minimize radio frequency emissions that could interfere with sensitive electronics.

Carburetor

The gasoline carburetor is a device that introduces fuel into the airstream as it flows into the engine. Gasoline is maintained in the float chamber by the float-actuated valve at a level slightly below the outlet of the jet. Air flows downward through the throat, past the throttle valve, and into the intake manifold. A throat is formed by the reduced diameter, and acceleration of the air through this smaller passage causes a decrease in pressure proportional to the amount of air flowing. This decrease in throat pressure results in fuel flow from the jet into the airstream. Any increase in airflow caused by change in engine speed or throttle position increases the pressure differential acting on the fuel and causes more fuel to flow.

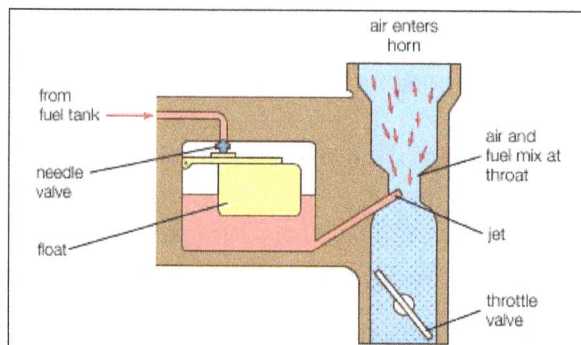

A simple carburetor.

The volume ratio of fuel to air established by the throat and fuel-jet sizes will be maintained with increased flow, but the weight ratio of fuel to air increases because the air expands to a lower density as the throat pressure decreases. This enriching tendency necessitates the inclusion of a compensating device in a practical carburetor. Carburetor design is further complicated by the need for an enriching device to provide a maximum-power ratio at full throttle, a choke to facilitate starting a cold engine, an idling system to provide the special needs of light-load operation, and an accelerating device to supply additional fuel while the throttle is being opened.

Fuel Injection

Most modern automobile engines use an electronic fuel-injection system in the intake manifold of the engine instead of a carburetor. The fuel-injection system is a closed-loop feedback system controlled by an engine management system that consists of sensors, an electric fuel pump, fuel injectors, fuel tubing, and valving. The engine management system controls both the ignition firing and the fuel management. In some designs the engine management system also controls the transmission. Sensors monitor the engine's operation and environmental conditions and transmit the data to the engine management system to determine how much fuel should be pumped to the fuel injectors for delivery to the engine. Typical sensors include the following: mass airflow, exhaust oxygen, engine revolutions per minute, manifold absolute pressure, barometric pressure, coolant temperature, throttle position, knock, vehicle speed, air-conditioning load, power steering load, crankshaft position, and camshaft position.

The principal advantages of gasoline injection over carburetors are improved fuel economy as a result of more-accurate fuel and air proportioning, greater power because of the elimination of fuel heating, elimination of inlet icing, and more-uniform and direct delivery of fuel load to the cylinders. Since fuel injection does not rely on an intake manifold vacuum to deliver fuel, electronic fuel injection is used with turbocharged engines.

Supercharger

The efficiency of the charging process in an automotive engine usually rises to a peak of slightly more than 80 percent at about half the rated speed of the engine and then decreases considerably at higher speed. This change in air charge per cycle with engine speed is reflected in proportionate changes in the torque, or turning effort, applied to the crankshaft and causes the power that the engine can deliver at full throttle to reach a maximum as engine speed increases. At speeds above this peaking speed, the air charge introduced per cycle falls off so rapidly that less power is developed than at lower speeds. The inability of the engine to draw in a full charge of fresh air at high speeds limits the power output of the engine.

Supercharging overcomes this disadvantage by using a pump or blower to raise the pressure of the air supplied to the cylinders and increase the weight of charge. The loss in power suffered by unsupercharged engines at high altitudes (e.g., flying or driving over mountains) can be largely restored. It is also possible to more than double the power of an engine by supercharging; however, increased charge density and temperature, resulting from supercharging, increase the tendency for combustion knock or roughness in the spark-ignition engine and thus necessitate an undesirable decrease in compression ratio or the use of an antiknock fuel.

The supercharging blower may be geared to the crankshaft, in which case the power consumed in driving it is added to the friction loss of the engine. A turbocharger employs a gas turbine operated by the exhaust gases to drive a centrifugal blower. The turbocharged engine not only gains increased power capacity but also operates at improved fuel economy. Historically, large airplane reciprocating gasoline engines were usually supercharged both by geared blowers and by turbochargers to provide the large pumping capacity needed at high altitude; however, these engines have generally been replaced by turboprop engines. High-performance general aviation aircraft typically use turbocharged engines.

Since compressing air prior to introducing it into the cylinder increases the charge-air temperature, the mass of air that can be introduced into the engine is less than that which would be possible if the compressed air were at ambient temperature. Consequently, engine charge-air coolers, commonly referred to as either intercoolers or aftercoolers, are used to reduce the temperature of the charge air. Both air-to-coolant and air-to-air type coolers are available.

Cooling System

The cylinders of internal-combustion engines require cooling because of the inability of the engine to convert all of the energy released by combustion into useful work. Liquid cooling is employed in most gasoline engines, whether the engines are for use in automobiles or elsewhere. The liquid is circulated around the cylinders to pick up heat and then through a radiator to dissipate the heat. Usually a thermostat is located in the circulating system to maintain the designed jacket temperature—approximately 88 °C (190 °F). The cooling system is usually pressurized to raise the boiling point of the coolant so that a higher outlet temperature can be maintained to improve thermal efficiency and increase the heat-transfer capacity of the radiator. A pressure cap on the radiator maintains this pressure by valves that open outwardly at the designed pressure and inwardly to prevent a vacuum as the system cools.

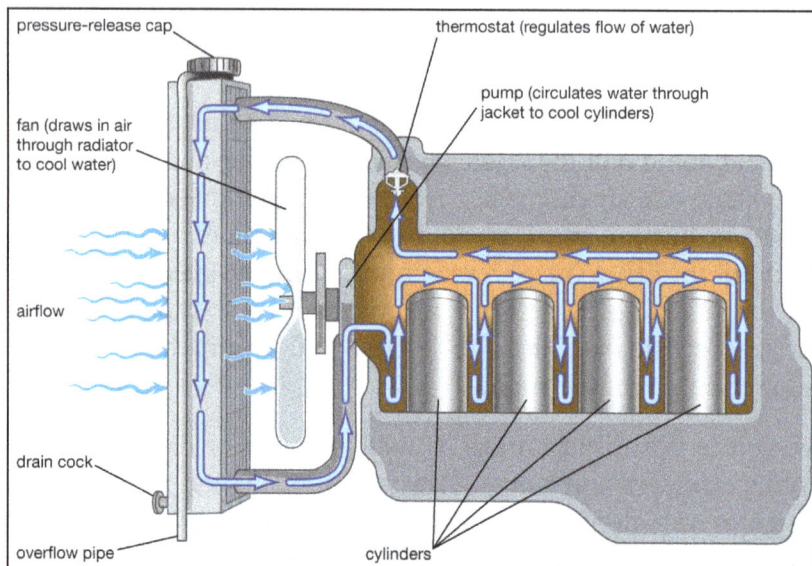

Typical gasoline engine cooling system.

Some engines, particularly aviation engines and small units for mowers, chain saws, and other tools, are air-cooled. Air cooling is accomplished by forming thin metal fins on the exterior surfaces of the cylinders to increase the rate of heat transfer by exposing more metal surface to the cooling air. Air is forced to flow rapidly through the spaces between the fins by ducting air toward the engine.

Lubrication System

Lubrication is employed to reduce friction by interposing a film between rubbing parts. The lubrication system must continuously replace the film.

The lubricants commonly employed are refined from crude oil after the fuels have been removed. Their viscosities must be appropriate for each engine, and the oil must be suitable for the severity of the operating conditions. Oils are improved with additives that reduce oxidation, inhibit corrosion, and act as detergents to disperse deposit-forming gums and solid contaminants. Motor oils also include an antifoaming agent. Various systems of numbers are used to designate oil viscosity; the lower the number, the lighter the body of the oil. Viscosity must be chosen to match the flow rate of oil through a part to the designed cooling requirements of the part. If the oil is too thick it will not flow through the part fast enough to properly dissipate heat. Certain oils contain additives that oppose their change in viscosity between winter and summer.

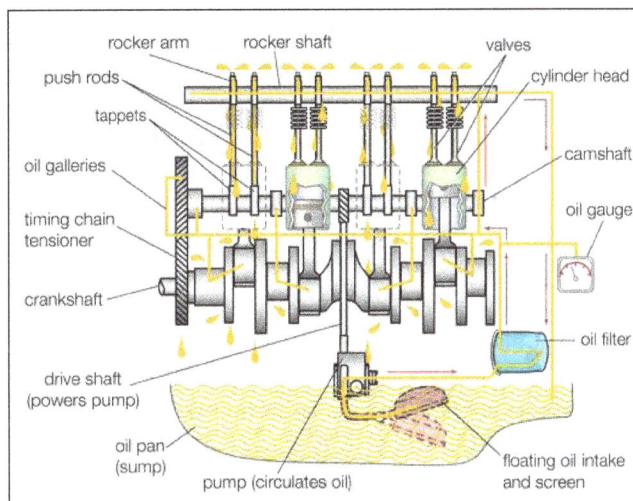

Typical gasoline engine lubrication system.

Oil filters, if regularly serviced, can remove solid contaminants from crankcase oil, but chemical reactions may form liquids that are corrosive and damaging. Depletion of the additives also limits the useful life of lubricating oils.

The lubrication system is fed by the oil sump that forms the lower enclosure of the engine. Oil is taken from the sump by a pump, usually of the gear type, and is passed through a filter and delivered under pressure to a system of passages or channels drilled through the engine. Virtually all modern engines use full-flow type oil filters. Filtered oil is supplied under pressure to crankshaft and camshaft main bearings. Adjacent crank throws are drilled to enable the oil to flow from the supply at the main bearings to the crankpins. Leaking oil from all of the crankshaft bearings is sprayed on the cylinder walls, cams, and up into the pistons to lubricate the piston pins. Additional passages intersect the cam-follower openings and supply oil to hydraulic valve lifters when used. A spring-loaded pressure-relief valve maintains the pressure at the proper level. Oil is important for both lubrication and cooling.

Exhaust System

Combustion products exit the engine cylinder through the exhaust valves in the cylinder head. Engines may be configured with either an exhaust manifold or an exhaust header. The exhaust manifold is a common chamber to which all the cylinders directly feed combustion products. The advantages of this method are manufacturing and positioning simplicity. The disadvantage is irregular backpressure at the exhaust ports of the cylinders. Headers are composed of a group of tubes, all of common length, connected on one end to each cylinder exhaust-valve location and on the other end to a common exit throat.

The exhaust gases in modern automotive engines next pass through an emission-control device. Emission-control sensors and catalytic converters for reducing air pollution are additional exhaust-system components. Typically, exhaust gases enter a catalytic converter to reduce nitric oxide emissions. The next chamber reduces unburned hydrocarbons and carbon monoxide exhaust emissions.

The reactor system for controlling emissions is often composed of a belt-driven air compressor connected to small nozzles installed in the exhaust manifold facing the outlet from each exhaust valve. A small jet of air is thus directed toward the red-hot outflowing combustion products to provide oxygen to consume the hydrocarbons and carbon monoxide. Sensors monitor exhaust-gas parameters (e.g., temperature and oxygen content) and, in electronic fuel-injection systems, provide information to the control unit to assist in reducing pollutant emissions.

Exhaust gases from an internal-combustion engine are passed through a muffler to suppress audible vibrations. When the exhaust valve opens, the pressure in the engine causes an initial gas outflow at explosive velocity. Successive discharges from the cylinders set up pressure pulsations that produce a sharp barking sound. The muffler damps out or absorbs these pulsations so that the gases leave the outlet as a relatively smooth, quiet stream.

Mufflers of early design contained sets of baffles that reversed the flow of the gases or otherwise caused them to follow devious paths so that interference between the pressure waves reduced the pulsations. The mufflers most commonly used in modern motor vehicles employ resonating chambers connected to the passages through which the gases flow. Gas vibrations are set up in each of these chambers at the fundamental frequency determined by its dimensions. These vibrations cancel or absorb those present in the exhaust stream of about the same frequency. Several such chambers, each tuned to one of the predominant frequencies present in the exhaust stream, effectively reduce noise.

Fuel

Gasoline was originally considered dangerous and was discarded and destroyed at early refineries, which were manufacturing kerosene for lamps. As the gasoline engine developed, gasoline and the engine were harmonized to attain the best possible matching of characteristics. The most important properties of gasoline are its volatility and antiknock quality. Volatility is a measure of the ease of vaporization of gasoline, which is adjusted in the production process to account for seasonal and altitude variations in the local market. Properly formulated gasoline helps engines to start in cold weather and to avoid vapour lock in hot weather.

To suit the needs of a modern engine, a gasoline must have the volatility for which the fuel system of the engine was designed and an antiknock quality sufficient to avoid knock under normal operation. Although other specifications must also be met, volatility and knock rating are the most important. The size and structural arrangement of the molecules principally determine the knocking tendency of a gasoline as well as its volatility.

Tetraethyl lead, added to gasolines for many years to improve antiknock fueling, has been found to contaminate the exhaust gases with poisonous lead oxides, and so the practice has ended. Lower compression ratios and improved combustion-chamber designs have eliminated the need for extremely high-antiknock gasolines.

Lubricating oil is added to gasoline used in crankcase-compression two-stroke-cycle engines.

Performance

The performance of an engine is expressed in terms of power, speed, and fuel economy. The three quantities are evaluated with a dynamometer, a laboratory device that applies a controllable load in the form of resistance to the turning of the crankshaft and also measures the torque exerted at the shaft coupling. The resistance imposed by a dynamometer may be adjusted so that the desired engine speed is established at any throttle position. It is thus possible to run the engine at various speeds throughout its operating range, to continuously maintain these operating conditions, and to measure the precise load and speed at which each run is made. Additional test equipment permits measurement of the exact quantity of fuel consumed, as well as the duration of the runs. From these data the power-speed-economy relationships can be calculated and performance plotted.

The power produced by an engine is expressed in horsepower. When the power developed is measured by means of a dynamometer or similar braking device, it is called brake horsepower. This is the power actually delivered by the engine and is therefore the capacity of the engine. The power developed in the combustion chambers of the engine is greater than the delivered power because of friction and other mechanical losses. This power loss, called the friction horsepower, can be evaluated by "motoring" the engine (driving it in a forward direction) with a suitable dynamometer when no fuel is being burned. The power developed in the cylinder can then be found by adding the friction horsepower to the brake horsepower. This quantity is the indicated horsepower of the engine, so called from an instrument known as the engine indicator, which is used to measure the pressure on the piston and thus calculate the power developed in the cylinder.

Mechanical efficiency is defined as brake horsepower in percent of indicated horsepower and is usually between 70 and 90 percent for normal operating speeds.

A quantity called brake mean effective pressure is obtained by multiplying the mean effective pressure of an engine by its mechanical efficiency. This is a commonly used index expressing the ability of the engine, per unit of cylinder bore, to develop both useful pressure in the cylinders and delivery power. If the power delivered is increased by any change other than an increase in speed or cylinder dimensions, its brake mean effective pressure increases proportionately.

Comparison with other Engines

When the gasoline engine is compared with other types of internal-combustion engines, certain similarities and differences, as well as some advantages and disadvantages, become apparent. The diesel engine and the gas engine (an engine utilizing a gas such as compressed natural gas or propane as the fuel) have a good deal in common with the gasoline engine, since they are all cylinder-and-piston engines that burn air-fuel mixtures in contact with moving components. The important difference that distinguishes the diesel engine is that it has no spark-ignition system. Compared with a gasoline engine of the same horsepower, the diesel engine is heavier and more expensive, but it has a longer life and operates at less cost per horsepower-hour because it burns less fuel.

The gas engine has much in common with the gasoline engine; in fact, in some instances their differences are very slight at best. Structurally, the difference lies primarily in the substitution of a gas-mixing valve for a carburetor. The cylinder and piston configurations are the same. In general, gases have better antiknock qualities than gasoline, permitting slightly higher compression ratios without knock or other combustion difficulties.

From the standpoint of application, the gas engine burning natural gas, manufactured gas, or industrial by-product gas is limited primarily to stationary power plant use because it must remain connected to the gas pipeline. If, however, the fuel is liquefied petroleum gas, sometimes called bottled gas, the containers of gas can be carried in a vehicle, leading to much flexibility in applications. The present obstacle is that facilities are not readily available for replenishing the gas supply.

Rotary Engines

Rotary engines or Wankel engines are a type of internal combustion engine, most popularly used in the Mazda RX-7, which converts heat from the combustion of a high pressure air/fuel mixture into useful work for the rest of the car. Its unique characteristic is its triangular rotor, which performs the same tasks as a reciprocating engine's piston would, but in a very different manner.

The rotor is contained in an oval shaped housing, and performs the common four-stroke cycle of an internal combustion engine, as seen in Figure. The rotor is connected to an output shaft which spins 3x faster than the rotor (inner circle labeled "B" in the Figure). This cycle is described below, and occurs 3 times for each spin of the rotor:

- Intake: This initiates when the tip of the rotor passes the intake port. At this moment, the chamber is at its smallest and as it rotates the chamber expands—drawing in the air/fuel mixture. As soon as the rotor end passes the intake port, it moves on to the compression stage, while the next face of the rotor starts this step over.

- Compression: As the rotor continues spinning, the air/fuel mixture becomes compressed because the chamber is decreasing in size. This is necessary for the next part, which ignites this mixture.

- Ignition: The compressed mixture gets ignited by spark plugs, and the vast increase in pressure forces the rotor to expand. This is the power-stroke, providing useful work. Two spark plugs are often needed to provide even ignition throughout the chamber. The exhaust gas expands into the chamber, until the rotor tip passes the exhaust port.

- Exhaust: Once the tip passes this port, the high pressure exhaust gases can flow through the exhaust port. The rotor continues to spin until the end of its face passes the exhaust port, the tip passes the intake port, and the cycle repeats.

The interesting part of this cycle is that every step is occurring at the same time, just in different chambers. This gives three power-strokes for every turn of the rotor.

Differences from a Reciprocating Engine

Besides the different method to complete the four-stroke cycle, rotary engines have different advantages and drawbacks from the more common reciprocating engines:

- Fewer moving parts: A two-rotor rotary engine has three moving parts—two rotors and an output shaft—while ordinary reciprocating engines have at least 40. This gives rotary engines better reliability.

- Smoother: The rotor spins constantly in one direction, unlike reciprocating engines whose pistons change direction abruptly. They are also counter-balanced by weights which reduce internal vibrations. The power delivery is also more continuous because of the three-power strokes for every turn of the rotor.

- Slower: The rotor spins at one-third the speed of the output shaft, so the main moving parts move slower than those in a piston engine. This improves reliability.

Drawbacks

Manufacturing costs can be higher due to the lower popularity of these engines. They also typically consume more fuel than other engines due to their low compression ratio and therefore have a lower thermal efficiency which makes it difficult for them to meet emission regulations.

V-Type Engines

An early Vee engine - this is a two-cylinder V-twin used in an early British motorcycle.

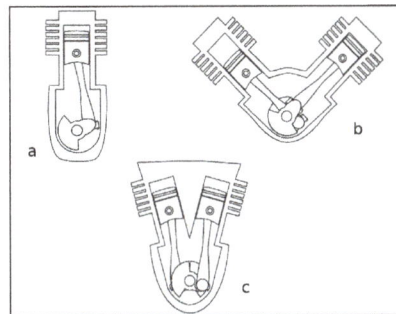

Three types of engine: a — straight engine, b — Vee engine, c — VR engine.

The yellow lines indicate the 'angle' of the 'Vee'.

A V engine, or Vee engine is a common configuration for an internal combustion engine. The cylinders and pistons are aligned, in two separate planes or 'banks', so that they appear to be in a "V" when viewed along the axis of the crankshaft. The Vee configuration generally reduces the overall engine length, height and weight compared with an equivalent inline configuration.

The first V-type engine, a 2-cylinder vee twin, was built in 1889 by Daimler, to a design by Wilhelm Maybach. By 1903 V8 engines were being produced for motor boat racing by the Société Antoinette to designs by Léon Levavasseur, building on experience gained with in-line four-cylinder engines. In 1904, the Putney Motor Works completed a new V12, 150bhp 18.4 litre engine – the first V12 engine produced for any purpose. This one was manufactured for two Russian brothers making a dirigible. They ran out of money and Commander May bought it on a sale or return basis for Motor boat racing, having some moderate success in 1908. The engine was exposed and the hot coil ignition created misfiring on becoming wet with the spray. Robert Bosch supplied the very first magnetos and the problem was solved.

Characteristics

Usually, each pair of corresponding pistons from each bank of cylinders share one crankpin on the crankshaft, either by master/slave rods or by two ordinary connecting rods side by side. However, some V-twin engine designs have two-pin cranks, while other V configurations include split crankpins for more even firing.

V-engines are generally more compact than straight engines with cylinders of the same dimensions and number. This effect increases with the number of cylinders in the engine; there might be no noticeable difference in overall size between V-twin and straight-twin engines while V8 engines are much more compact than straight-eight engines.

Various cylinder bank angles of Vee are used in different engines; depending on the number of cylinders, there may be angles that work better than others for stability. Very narrow angles of Vee combine some of the advantages of the Vee engine and the straight engine (primarily in the form of compactness) as well as disadvantages; the concept is an old one pioneered by Lancia's V4 engine in the 1920s, but recently reworked by Volkswagen Group with their VR engines, which is actually a combination of V and straight configuration.

Some Vee configurations are well-balanced and smooth, while others are less smoothly running than their equivalent straight counterparts. V8s with crossplane crankshaft can be easily balanced with the use of counterweights only. V12s, being in effect two straight-6 engines married together, are fully balanced; if the V-angle is 60° for 4-stroke or 30° for 2-stroke, they also have even firing. Others, such as the V2, V4, V6, flatplane V8, V10, V14 and V18 engine show increased vibration and generally require balance shafts or split crankshafts.

Inverted Engines

Certain types of Vee engine have been built as inverted engines, most commonly for aircraft. Advantages include better visibility in a single-engined airplane, and lower centre of gravity. Examples include World War II German Daimler-Benz DB 601, Junkers Jumo, and Argus Motoren piston engines.

Flat Engines

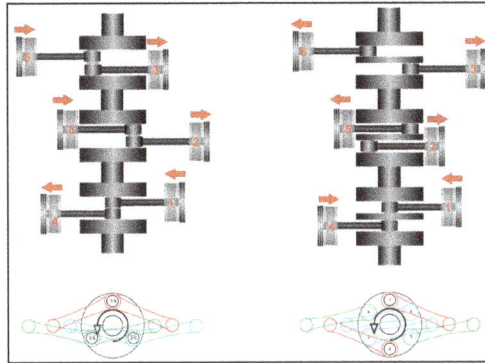

Difference between two flat 6 cylinder engines: 180° V on the left, boxer on the right.

A flat engine is an internal combustion engine with horizontally-opposed cylinders. Typically, the layout has cylinders arranged in two banks on either side of a single crankshaft and is otherwise known as the boxer, or horizontally-opposed engine. The concept was patented in 1896 by engineer Karl Benz, who called it the "contra engine."

A boxer engine should not be confused with the opposed-piston engine, in which each cylinder has two pistons but no cylinder head. Also, if a straight engine is canted 90 degrees into the horizontal plane, it may be thought of as a "flat engine". Horizontal inline engines are quite common in industrial applications such as underfloor mounting for buses.

True boxers have each crankpin controlling only one piston/cylinder while the 180° engines, which superficially appear very similar, share crankpins. The 180° engine, which may be thought of as a type of V engine, is quite uncommon as it has all of the disadvantages of a flat engine, and few of the advantages.

Boxer Engine

1954 BMW R68 engine. The two cylinders are offset.

In 1896, Karl Benz invented the first internal combustion engine with horizontally opposed pistons. He called it the *kontra* engine, as the action of each side opposed the action of the other. This design has since been called the "boxer" engine because each pair of pistons moves in and out together, rather like the gloves of a boxer. The boxer engine has pairs of pistons reaching TDC simultaneously.

The boxer configuration is the only configuration in common use that does not have unbalanced forces with a four-stroke cycle regardless of the number of cylinders, as long as both banks have the same number of cylinders. These engines do not require a balance shaft or counterweights on the crankshaft to balance the weight of the reciprocating parts, which are required in most other engine configurations. However, in the case of boxer engines with fewer than six cylinders, unbalanced moments (a reciprocating torque also known as a "rocking couple") are unavoidable due to the "opposite" cylinders being slightly out of line with each other. Other engine configurations with natural dynamic balance include the straight-six, the straight-eight, the V12, and the V16.

Boxer engines (and flat engines in general) tend to be noisier than other common engines for both intrinsic and other reasons. In cars, valve clatter from the engine compartment is not damped by air filters or other components.

Aviation Use

UL260i flat-4 aircraft engine.

In 1909 Santos Dumont used Dutheil-Chalmers and Darracq boxer engines in his Demoiselle airplane, the first airplane with significant production (over 40).

Multi-cylinder boxer layouts have proved to be well suited as light aircraft engines, as exemplified by Continental, Lycoming, Rotax, Jabiru and Verner. An important factor in aircraft use is the flat engine's absence of vibration, which allows a lighter engine mount.

General aviation aircraft often use air-cooled flat-four and flat-six engines made by companies such as Lycoming and Continental. Ultralight and microlight aircraft often use engines such as the Rotax 912 and Jabiru 2200.

During the Second World War, Boxer engines were used as a starter motor for the first German jet engines to power up the engine at cranking speed. The two-cylinder two-stroke flat engine was developed by Norbert Riedel ("Riedel starter"), had a cylinder capacity of 270 cc and a power of 8 kW (10.5 hp) at 7150 rpm and essentially functioned as a pioneering example of an APU for starting a jet engine. It was an extreme short stroke (bore / stroke: 70 mm / 35 mm = 2:1) design so it could fit in the hub of the turbine compressor and started electrically or with a pull starter. The engine was produced by the Victoria works in Nuremberg and served as a starter for the jet engines Junkers Jumo 004 and BMW 003.

Riedel starter for German WWII jet engines, with pull-start handle and cable.

Motorcycle Use

Flat engines offer several advantages for motorcycles, namely: a low centre of gravity, smoothness, suitability for shaft drive, and (if air-cooled) excellent cooling of the cylinders.

The first motorcycle with a boxer engine was the 1905 Fée flat-twin, which was developed into the 1907 Douglas. Douglas would continue making flat-twin motorcycles until 1957, ending with the Dragonfly. BMW have made motorcycles with flat-twin engines since the BMW R32 of 1923. Unlike contemporary Douglas motorcycles, which had their engines mounted with the cylinders in line with the frame, the R32 had its cylinders mounted across the frame and used a shaft to drive the rear wheel. The drivetrain layout of the R32 has been used, with improvements, in all subsequent BMW flat-twin motorcycles. The Russian Ural and Ukrainian Dnepr flat twins were copies of the pre-WWII military plunger-suspension BMW R71.

In 1923, Max Friz designed the first BMW motorcycles, choosing a 500 cc boxer engine and unit transmission with shaft drive. This engine type is still in production today. The BMW 247 engine, known as an airhead due to its air cooling, was produced until 1995. BMW replaced it with the oilhead engine with partial oil cooling and four valves per cylinder, but still retained the same flat-twin configuration. In 2013, BMW introduced partial ("precision") water-cooled version, first on their BMW GS and planned to replace all oilheads.

Flat-four engines have been used in the 1938–1939 Zündapp K800, the French BFG motorcycle with the Citroën GS engine, and the Honda Gold Wing from 1975 to 1986. Gold Wings since 1987 have used flat-six engines, as have Honda Valkyries.

Automotive Use

The low centre of gravity allowed by a flat engine can reduce body roll in automobiles and enhance handling precision. Historically they could be found in cars manufactured by companies such as Porsche, Lancia, Benz, Ford, Tatra, Citroen, Alfa Romeo, Jowett, Rover, Volkswagen, Chevrolet, and Ferrari. The most prominent manufacturers currently using a boxer engine as their primary engine configuration are Porsche and Subaru.

Hino Motors DS140 12-cylinder boxer diesel engine.

Automobile Layouts used with Flat Engines

Citroën 2CV drivetrain, including overhanging front-mounted flat-twin engine
and shafts to front wheels, in a Blackjack Avion three-wheeler.

When mounted longitudinally in a vehicle, flat engines with up to six cylinders are short, low, and wide. As a result, they have often been used in compact drivetrains where the engine is mounted outside the wheelbase and drives the nearer pair of wheels through the transmission without a drive shaft. The short length of a longitudinally mounted flat engine with six cylinders or fewer makes it ideally suitable for air cooling.

Examples with rear-engine, rear-wheel-drive layouts include the two-cylinder BMW 600 and 700, four-cylinder Tatra 97, Volkswagen Beetle and Porsche 356, and the six-cylinder Chevrolet Corvair, Porsche 911, and Tucker 48. All of these examples except the Tucker and later versions of the Porsche 911 are air-cooled.

Tatra 11 backbone chassis with front-mounted flat-twin engine and rear final drive.

Examples with front-engine, front-wheel-drive layouts / four-wheel drive layouts include two-cylinder Citroëns and Panhards and the four-cylinder Citroën GS, Lancias from the Flavia to the Gamma, Alfa Romeo Alfasud and Subarus DL and GL. The Citroëns and Panhards are air cooled while the Lancias, Alfa Romeos and Subarus are water cooled.

Flat engines have also been used in cars with front-engine, rear-wheel-drive layouts, including Bradford trucks and vans, the Glas Isar, Jowett cars and trucks, early Tatras, the Scion FR-S and the Subaru BRZ.

Flat engines, including non-boxer flat-12s, have been mid-mounted in Porsche and Ferrari racing cars. Porsche has made the 914 for road use with four or six cylinder air cooled boxer engines, while Ferrari road cars with mid-mounted water cooled non-boxer flat-12s include the 365 GT4 BB, BB 512, BB 512i, Testarossa, 512 TR, and F512 M.

Subaru use a four-cylinder flat engine at the front of the car that drives all four wheels. The front half-shafts come out of a front differential that is part of the gearbox. A rear driveshaft connects the gearbox to the rear half-shafts.

Current Users

Porsche

All early Porsches up to the 914 used air-cooled boxer engines. The 356 and its derivatives had flat-four engines, as did the 912 and the mid-engined 914. The original 911 had an air-cooled flat-six, as did a six-cylinder version of the 914; the later 964 and 993 versions of the 911 used air/oil cooled flat-six engines.

Some Porsche sports racing models, including the 908, used flat-eight engines. The flat-twelve in the 917 model is a 180° V 12 engine; Ferrari's Berlinetta Boxer (BB 512) later used a 180-degree flat 12 for Le Mans to compete with Porsche, but Ferrari also used the BB 512 on road-going cars, unlike Porsche.

Porsche began to use water-cooled flat-six engines in the Boxster in 1996. This type is also used in the Cayman and in 911 models starting with the 996. The 718 Boxster and Cayman have moved from naturally aspirated flat sixes, to turbocharged flat fours, while the 911 uses a smaller turbocharged flat six.

Subaru

Since the Subaru 1000 of 1966 the water-cooled front-mounted flat-four and flat-six engines have been used by Subaru in all of its mid-sized cars, including the Impreza, Legacy, Outback, and SVX. The Forester and Tribeca SUVs, BRAT and Baja pickup trucks, and BRZ sports car also use boxer engines. Subaru refers to these as boxer engines in publicity commentary, and include a variety of naturally aspirated and turbo driven engines from 1966, when the Subaru 1000 was introduced to current; both closed and semi-closed short blocks have been used. A print ad for the 1973 Subaru GL coupe referred to the engine as "quadrozontal" The EJ series of four-cylinder engines released first in 1990 has been the focus for the development of the boxer engine in the late 20th century. Ranging from 1.5 to 2.5 litres, this engine in its 2-litre turbo arrangement has been the power behind World Rally Championship winning cars. Subaru also offers a common rail boxer turbo-diesel, called the Subaru EE series, the world's first to be fitted into a passenger car.

Subaru boxer turbodiesel engine cutaway display.

Toyota

In a joint venture between Subaru and Toyota, a 1,998 cc flat-four engine with 200 PS (147 kW) having GDI was developed as the Subaru FA20 and Toyota 4U-GSE. This was used in the two-door coupe Toyota 86 and Subaru BRZ.

W Motors

The Lykan HyperSport, the third most expensive car in the world as of 2015, uses a Porsche-derived flat-six engine.

Straight Engine

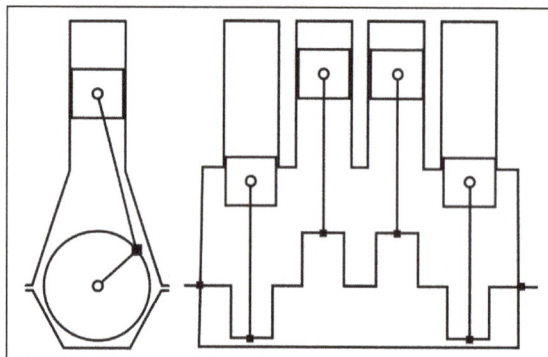

4-cylinder straight engine scheme.

The straight or inline engine is an internal-combustion engine with all cylinders aligned in one row and having no offset. Usually found in four, six and eight cylinder configurations, they have been used in automobiles, locomotives and aircraft, although the term in-line has a broader meaning when applied to aircraft engines.

A straight engine is considerably easier to build than an otherwise equivalent horizontally opposed or V engine, because both the cylinder bank and crankshaft can be milled from a single metal casting, and it requires fewer cylinder heads and camshafts. In-line engines are also smaller in overall physical dimensions than designs such as the radial, and can be mounted in any direction. Straight configurations are simpler than their V-shaped counterparts. Although six-cylinder engines are inherently balanced, the four-cylinder models are inherently off balance and rough, unlike 90-degree V fours and horizontally opposed 'boxer' four cylinders.

Automobile use

The inline-four engine is the most common four-cylinder configuration, whereas the straight-6 has largely given way to the V6 engine, which although not as naturally smooth-running is smaller in both length and height and easier to fit into the engine bay of smaller modern cars. Some manufacturers, including Acura, Audi, Ford, Mercedes-Benz, Volkswagen and Volvo, have also used straight-five configurations. The General Motors *Atlas* family includes straight-four, straight-five, and straight-six engines. Some small cars have inline three engines.

Once, the straight-eight was the prestige engine arrangement; it could be made more cheaply than a V-engine by luxury car makers, who would focus on other specifics than the geometric ones, and even built engines more powerful than any V8 engine. In the 1930s, Duesenberg used a cylinder block made from aluminium alloy, with four valves per cylinder and hemispherical heads to produce the most powerful engine on the market. It was thus a selling point for Pontiac to introduce the cheapest straight-eight in 1933. However, following World War II, the straight-eight was supplanted by the lighter and more compact V8 engine, which allowed shorter engine bays to be used in the design.

When a straight engine is mounted at an angle from the vertical it is called a slant engine. Chrysler's Slant 6 was used in many models in the 1960s and 1970s. Honda also often mounts its straight-four and straight-five engines at a slant, as on the Honda S2000 and Acura Vigor. SAAB initially used the Triumph Slant-4 engine tilted at 45 degrees for the Saab 99, but later versions of the engine were less tilted.

Two main factors have led to the recent decline of the straight-six in automotive applications. First, Lanchester balance shafts, an old idea reintroduced by Mitsubishi in the 1980s to overcome the natural imbalance of the inline-four engine and rapidly adopted by many other manufacturers, have made both inline-four and V6 engines smoother-running; the greater smoothness of the straight-six layout is no longer as great an advantage. Second, fuel consumption became more important, as cars became smaller and more space-efficient. The engine bay of a modern small or medium car, typically designed for an inline-four, often does not have room for a straight-six, but can fit a V6 with only minor modifications.

Some manufacturers (originally Lancia, and more recently Volkswagen with the VR6 engine) have attempted to combine advantages of the straight and V configurations by producing a narrow-angle V; this is more compact than either configuration, but is less smooth (without balancing) than either.

Straight-6 engines are used in some models from BMW, Ford, Jeep, Chevrolet, GMC, Toyota, Suzuki and Volvo Cars.

Bus and Rail Use

Some buses and trains with straight engines have their engines mounted with the row of cylinders horizontal. This differs from a flat engine because it is essentially an inline engine laid on its side. Underfloor engines for buses and diesel multiple units (DMUs) commonly use this design. Such engines may be based on a conventional upright engine with alterations to make it suitable for horizontal mounting.

Aviation Use

de Havilland Gipsy Major engine, an inverted inline-4 engine,
mounted in a de Havilland Australia DHA-3 Drover.

In aviation, the term "inline engine" is used more broadly, for any non-radial reciprocating engine.

Many straight engines, in the strict sense, have been produced for aircraft, particularly from the early years of aviation through the interwar period. Straight engines were simple and had low frontal area, reducing drag. Straight sixes were especially popular in World War One, including examples like the Mercedes D.III and the Siddeley Puma.

Some straight aircraft engines have been inverted rather than upright engines. Renault produced a straight-four inverted air-cooled motor, which was used on the Stampe SV.4. A similar design was the de Havilland Gipsy series of engines, used on the Tiger Moth and other aircraft. Advantages of the inverted arrangement include improved visibility for the pilot in single-engined craft, and lower center of gravity.

Motorcycle Use

Straight-4 engine installed in line with the frame of an Indian Four motorcycle.

In motorcycling, the term "in-line" is sometimes used narrowly, for a straight engine mounted in line with the frame. A two-cylinder straight engine mounted across the frame is sometimes called a parallel twin. Other times, motorcycling experts treat the terms parallel, straight, and inline as equivalent, and use them interchangeably.

Diesel Engines

Diesel engine built by Langen & Wolf under licence.

Shell Oil film showing the development of the diesel engine.

The diesel engine (also known as a compression-ignition or CI engine), named after Rudolf Diesel, is an internal combustion engine in which ignition of the fuel is caused by the elevated temperature of the air in the cylinder due to the mechanical compression (adiabatic compression). This contrasts with spark-ignition engines such as a petrol engine (gasoline engine) or gas engine (using a gaseous fuel as opposed to petrol), which use a spark plug to ignite an air-fuel mixture.

Diesel engines work by compressing only the air. This increases the air temperature inside the cylinder to such a high degree that atomised diesel fuel injected into the combustion chamber ignites spontaneously. With the fuel being injected into the air just before combustion, the dispersion of the fuel is uneven; this is called a heterogeneous air-fuel mixture. The torque a diesel engine produces is controlled by manipulating the air ratio; instead of throttling the intake air, the diesel engine relies on altering the amount of fuel that is injected, and the air ratio is usually high.

The diesel engine has the highest thermal efficiency (engine efficiency) of any practical internal or external combustion engine due to its very high expansion ratio and inherent lean burn which enables heat dissipation by the excess air. A small efficiency loss is also avoided compared to two-stroke non-direct-injection gasoline engines since unburned fuel is not present at valve overlap and therefore no fuel goes directly from the intake/injection to the exhaust. Low-speed diesel engines (as used in ships and other applications where overall engine weight is relatively unimportant) can reach effective efficiencies of up to 55%.

Diesel engines may be designed as either two-stroke or four-stroke cycles. They were originally used as a more efficient replacement for stationary steam engines. Since the 1910s they have been used in submarines and ships. Use in locomotives, trucks, heavy equipment and electricity generation plants followed later. In the 1930s, they slowly began to be used in a few automobiles. Since the 1970s, the use of diesel engines in larger on-road and off-road vehicles in the US has increased. According to Konrad Reif, the EU average for diesel cars accounts for half of newly registered cars.

The world's largest diesel engines put in service are 14-cylinder, two-stroke watercraft diesel engines; they produce a peak power of almost 100 MW each.

Operating Principle

Characteristics

The characteristics of a diesel engine are:

- Compression ignition: Due to almost adiabatic compression, the fuel ignites without any ignition-initiating apparatus such as spark plugs.

- Mixture formation inside the combustion chamber: Air and fuel are mixed in the combustion chamber and not in the inlet manifold.

- Engine speed adjustment solely by mixture quality: Instead of throttling the air-fuel mixture, the amount of torque produced (resulting in crankshaft rotational speed differences) is set solely by the mass of injected fuel, always mixed with as much air as possible.

- Heterogeneous air-fuel mixture: The dispersion of air and fuel in the combustion chamber is uneven.

- High air ratio: Due to always running on as much air as possible and not depending on exact mixture of air and fuel, diesel engines have an air-fuel ratio leaner than stochiometric.

- Diffusion flame: At combustion, oxygen first has to defuse into the flame, rather than having oxygen and fuel already mixed before combustion, which would result in a premixed flame.

- Fuel with high ignition performance: As diesel engines solely rely on compression ignition, fuel with high ignition performance (cetane rating) is ideal for proper engine operation, fuel with a good knocking resistance (octane rating), e.g. petrol, is suboptimal for diesel engines.

Cycle of the Diesel Engine

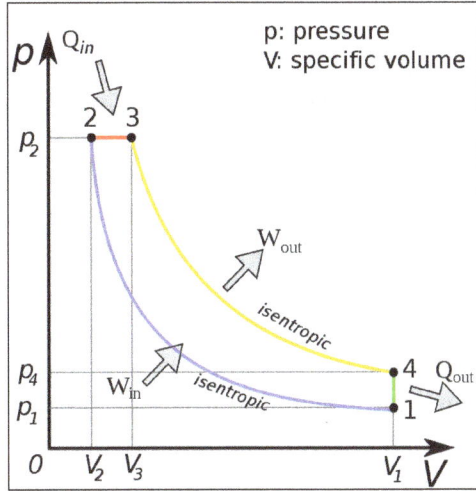

p-V Diagram for the ideal diesel cycle. The cycle follows the numbers 1–4 in clockwise direction. The horizontal axis is volume of the cylinder. In the diesel cycle the combustion occurs at almost constant pressure. On this diagram the work that is generated for each cycle corresponds to the area within the loop.

Diesel engine model, left side.

Diesel engine model, right side.

The diesel internal combustion engine differs from the gasoline powered Otto cycle by using highly compressed hot air to ignite the fuel rather than using a spark plug (*compression ignition* rather than *spark ignition*).

In the diesel engine, only air is initially introduced into the combustion chamber. The air is then compressed with a compression ratio typically between 15:1 and 23:1. This high compression causes the temperature of the air to rise. At about the top of the compression stroke, fuel is injected directly into the compressed air in the combustion chamber. This may be into a (typically toroidal) void in the top of the piston or a *pre-chamber* depending upon the design of the engine. The fuel injector ensures that the fuel is broken down into small droplets, and that the fuel is distributed evenly. The heat of the compressed air vaporises fuel from the surface of the droplets. The vapour is then ignited by the heat from the compressed air in the combustion chamber, the droplets continue to vaporise from their surfaces and burn, getting smaller, until all the fuel in the droplets has been burnt. Combustion occurs at a substantially constant pressure during the initial part of the power stroke. The start of vaporisation causes a delay before ignition and the characteristic diesel knocking sound as the vapour reaches ignition temperature and causes an abrupt increase in pressure above the piston (not shown on the P-V indicator diagram). When combustion is complete the combustion gases expand as the piston descends further; the high pressure in the cylinder drives the piston downward, supplying power to the crankshaft.

As well as the high level of compression allowing combustion to take place without a separate ignition system, a high compression ratio greatly increases the engine's efficiency. Increasing the compression ratio in a spark-ignition engine where fuel and air are mixed before entry to the cylinder is limited by the need to prevent damaging pre-ignition. Since only air is compressed in a diesel engine, and fuel is not introduced into the cylinder until shortly before top dead centre (TDC), premature detonation is not a problem and compression ratios are much higher.

The p–V diagram is a simplified and idealised representation of the events involved in a diesel engine cycle, arranged to illustrate the similarity with a Carnot cycle. Starting at 1, the piston is at bottom dead centre and both valves are closed at the start of the compression stroke; the cylinder contains air at atmospheric pressure. Between 1 and 2 the air is compressed adiabatically – that is without heat transfer to or from the environment – by the rising piston. (This is only approximately true since there will be some heat exchange with the cylinder walls.) During this compression, the volume is reduced, the pressure and temperature both rise. At or slightly before 2 (TDC) fuel is injected and burns in the compressed hot air. Chemical energy is released and this constitutes an injection of thermal energy (heat) into the compressed gas. Combustion and heating occur between 2 and 3. In this interval the pressure remains constant since the piston descends, and the volume increases; the temperature rises as a consequence of the energy of combustion. At 3 fuel injection and combustion are complete, and the cylinder contains gas at a higher temperature than at 2. Between 3 and 4 this hot gas expands, again approximately adiabatically. Work is done on the system to which the engine is connected. During this expansion phase the volume of the gas rises, and its temperature and pressure both fall. At 4 the exhaust valve opens, and the pressure falls abruptly to atmospheric (approximately). This is unresisted expansion and no useful work is done by it. Ideally the adiabatic expansion should continue, extending the line 3–4 to the right until the pressure falls to that of the surrounding air, but the loss of efficiency caused by this unresisted expansion is justified by the practical difficulties involved in recovering it (the engine would have to be much larger). After the opening of the exhaust valve, the exhaust stroke follows, but this (and the following induction stroke) are not shown on the diagram. If shown, they would be represented by a low-pressure loop at the bottom of the diagram. At 1 it is assumed that the exhaust and induction strokes have been completed, and the cylinder is again filled with air. The piston-cylinder system absorbs energy between 1 and 2 – this is the work needed to compress the air in the cylinder, and is provided by mechanical kinetic energy stored in the flywheel of the engine. Work output is done by the piston-cylinder combination between 2 and 4. The difference between these two increments of work is the indicated work output per cycle, and is represented by the area enclosed by the p–V loop. The adiabatic expansion is in a higher pressure range than that of the compression because the gas in the cylinder is hotter during expansion than during compression. It is for this reason that the loop has a finite area, and the net output of work during a cycle is positive.

Efficiency

Due to its high compression ratio, the diesel engine has a high efficiency, and the lack of a throttle valve means that the charge-exchange losses are fairly low, resulting in a low specific fuel consumption, especially in medium and low load situations. This makes the diesel engine very economical. Even though diesel engines have a theoretical efficiency of 75%, in practice it is much lower. In his 1893 essay *Theory and Construction of a Rational Heat Motor*, Rudolf Diesel describes that the effective efficiency of the diesel engine would be in between 43.2% and 50.4% , or maybe even greater. Modern passenger car diesel engines may have an effective efficiency of up to 43% , whilst engines in large diesel trucks, and buses can achieve peak efficiencies around 45%. However, average efficiency over a driving cycle is lower than peak efficiency. For example, it might be 37% for an engine with a peak efficiency of 44%. The highest diesel engine efficiency of up to 55% is achieved by large two-stroke watercraft diesel engines.

Major Advantages

Diesel engines have several advantages over engines operating on other principles:

- The diesel engine has the highest effective efficiency of all combustion engines.

  - Diesel engines inject the fuel directly into the combustion chamber, have no intake air restrictions apart from air filters and intake plumbing and have no intake manifold vacuum to add parasitic load and pumping losses resulting from the pistons being pulled downward against intake system vacuum. Cylinder filling with atmospheric air is aided and volumetric efficiency is increased for the same reason.

  - Although the fuel efficiency (mass burned per energy produced) of a diesel engine drops at lower loads, it doesn't drop quite as fast as that of a typical petrol or turbine engine.

Bus powered by biodiesel.

Diesel engines can combust a huge variety of fuels, including several fuel oils, that have advantages over fuels such as petrol. These advantages include:

- Low fuel costs, as fuel oils are relatively cheap.

- Good lubrication properties.

- High energy density.

- Low risk of catching fire, as they do not form a flammable vapour.

- Biodiesel is an easily synthesised, non-petroleum-based fuel (through transesterification) which can run directly in many diesel engines, while gasoline engines either need adaptation to run synthetic fuels or else use them as an additive to gasoline (e.g., ethanol added to gasohol).

- Diesel engines have a very good exhaust-emission behaviour. The exhaust contains minimal amounts of carbon monoxide and hydrocarbons. Direct injected diesel engines emit approximately as much nitrogen oxide as Otto cycle engines. Swirl chamber and precombustion chamber injected engines, however, emit approximately 50% less nitrogen oxide than Otto cycle engines when running under full load. Compared with Otto cycle engines, diesel engines emit 10 times less pollutants and 3 times less carbon dioxide.

- They have no high voltage electrical ignition system, resulting in high reliability and easy adaptation to damp environments. The absence of coils, spark plug wires, etc., also eliminates a source of radio frequency emissions which can interfere with navigation and communication equipment, which is especially important in marine and aircraft applications, and for preventing interference with radio telescopes. (For this reason, only diesel-powered vehicles are allowed in parts of the American National Radio Quiet Zone.)

- Diesel engines can accept super- or turbocharging pressure without any natural limit, constrained only by the design and operating limits of engine components, such as pressure, speed and load. This is unlike petrol engines, which inevitably suffer detonation at higher pressure if engine tuning and/or fuel octane adjustments are not made to compensate.

Fuel Injection

Diesel engines rely on internal mixture formation, which means that they require a fuel injection system. The fuel is injected directly into the combustion chamber, which can be either a segmented combustion chamber or an unsegmented combustion chamber. Fuel injection with the latter is referred to as *direct injection* (DI), whilst injection into the former is called *indirect injection* (IDI). In diesel engine terminology, indirect injection does not mean fuel injection into the inlet manifold or anywhere else outside the cylinder or combustion chamber: in fact, the definition of the diesel engine excludes such injection methods. For creating the fuel pressure, diesel engines usually have an injection pump. There are several different types of injection pumps and methods for creating a fine air-fuel mixture. Over the years many different injection methods have been used. These can be described as the following:

- Air blast, where the fuel is blown into the cylinder by a blast of air.

- Solid fuel / hydraulic injection, where the fuel is pushed through a spring loaded valve / injector to produce a combustible mist.

- Mechanical unit injector, where the injector is directly operated by a cam and fuel quantity is controlled by a rack or lever.

- Mechanical electronic unit injector, where the injector is operated by a cam and fuel quantity is controlled electronically.

- Common rail mechanical injection, where fuel is at high pressure in a common rail and controlled by mechanical means.

- Common rail electronic injection, where fuel is at high pressure in a common rail and controlled electronically.

Torque Controlling

Due to the way diesel engines work, a vital component of all diesel engines is a mechanical or electronic governor which regulates the torque of the engine and thus idling speed and maximum speed by controlling the rate of fuel delivery. This means a change of λ_v. Unlike Otto-cycle engines, incoming air is not throttled. Mechanically governed fuel injection systems are driven by the engine's gear train. These systems use a combination of springs and weights to control fuel delivery relative to both load and speed. Modern electronically controlled diesel engines control fuel delivery by use of an electronic control module (ECM) or electronic control unit (ECU). The ECM/ECU receives an engine speed signal, as well as other operating parameters such as intake manifold pressure and fuel temperature, from a sensor and controls the amount of fuel and start of injection timing through actuators to maximise power and efficiency and minimise emissions. Controlling the timing of the start of injection of fuel into the cylinder is a key to minimizing emissions, and maximizing fuel economy (efficiency), of the engine. The timing is measured in degrees of crank angle of the piston before top dead centre. For example, if the ECM/ECU initiates fuel injection when the piston is 10° before TDC, the start of injection, or timing, is said to be 10° before TDC. Optimal timing will depend on the engine design as well as its speed and load.

Types of Fuel Injection

Air-blast Injection

Typical early 20th century air-blast injected diesel engine, rated at 59 kW.

Diesel's original engine injected fuel with the assistance of compressed air, which atomised the fuel and forced it into the engine through a nozzle (a similar principle to an aerosol spray). The nozzle opening was closed by a pin valve lifted by the camshaft to initiate the fuel injection before top dead centre (TDC). This is called an air-blast injection. Driving the compressor used some power but the efficiency was better than the efficiency of any other combustion engine at that time. Also, air-blast injection made engines very clunky and heavy and did not allow for quick load alteration, thus rendering it unusable for road vehicles.

Indirect Injection

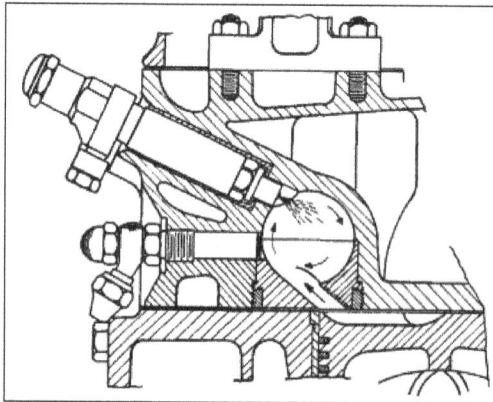

Ricardo Comet indirect injection chamber.

An indirect diesel injection system (IDI) engine delivers fuel into a small chamber called a swirl chamber, precombustion chamber, pre chamber or ante-chamber, which is connected to the cylinder by a narrow air passage. Generally the goal of the pre chamber is to create increased turbulence for better air / fuel mixing. This system also allows for a smoother, quieter running engine, and because fuel mixing is assisted by turbulence, injector pressures can be lower. Most IDI systems use a single orifice injector. The pre-chamber has the disadvantage of lowering efficiency due to increased heat loss to the engine's cooling system, restricting the combustion burn, thus reducing the efficiency by 5–10%. IDI engines are also more difficult to start and usually require the use of glow plugs. IDI engines may be cheaper to build but generally require a higher compression ratio than the DI counterpart. IDI also makes it easier to produce smooth, quieter running engines with a simple mechanical injection system since exact injection timing is not as critical. Most modern automotive engines are DI which have the benefits of greater efficiency and easier starting; however, IDI engines can still be found in the many ATV and small diesel applications. Indirect injected diesel engines use pintle-type fuel injectiors.

Helix-controlled Direct Injection

Direct injection Diesel engines inject fuel directly into the cylinder. Usually there is a combustion cup in the top of the piston where the fuel is sprayed. Many different methods of injection can be used. Usually, an engine with helix-controlled mechanic direct injection has either an inline or a distributor injection pump. For each engine cylinder, the corresponding plunger in the fuel pump measures out the correct amount of fuel and determines the timing of each injection. These engines use injectors that are very precise spring-loaded valves that open and close at a specific fuel pressure. Separate high-pressure fuel lines connect the fuel pump with each cylinder. Fuel volume for each single combustion is controlled by a slanted groove in the plunger which rotates only a few degrees releasing the pressure and is controlled by a mechanical governor, consisting of weights rotating at engine speed constrained by springs and a lever. The injectors are held open by the fuel pressure. On high-speed engines the plunger pumps are together in one unit. The length of fuel lines from the pump to each injector is normally the same for each cylinder in order to obtain the same pressure delay. Direct injected diesel engines usually use orifice-type fuel injectors.

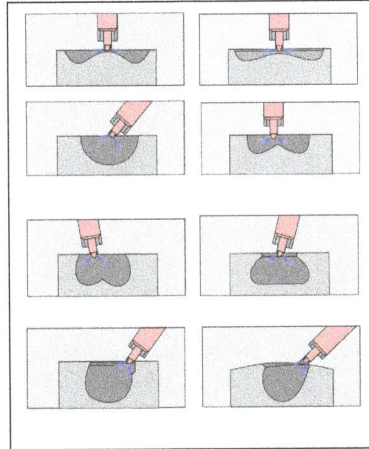

Different types of piston bowls.

Electronic control of the fuel injection transformed the direct injection engine by allowing much greater control over the combustion.

Unit Direct Injection

Unit direct injection, also known as Pumpe-Düse (*pump-nozzle*), is a high pressure fuel injection system that injects fuel directly into the cylinder of the engine. In this system the injector and the pump are combined into one unit positioned over each cylinder controlled by the camshaft. Each cylinder has its own unit eliminating the high-pressure fuel lines, achieving a more consistent injection. Under full load, the injection pressure can reach up to 220 MPa. Unit injection systems used to dominate the commercial diesel engine market, but due to higher requirements of the flexibility of the injection system, they have been rendered obsolete by the more advanced common-rail-system.

Common Rail Direct Injection

Common rail (CR) direct injection systems, unlike other injection systems, do not have a combined pressure creation and injection apparatus. A high-pressure injection pump creates a constant pressure, not depending on the engine speed or fuel mass injected. A buffer, the so-called rail, saves this pressure. This allows fuel injection at any given moment, even multiple injections in a very short amount of time. The Electronic Diesel Control unit (EDC) controls both rail pressure and injections depending on several different parameters of the engine. The injectors of older CR systems have solenoid-driven plungers for lifting the injection needle, whilst newer CR injectors use plungers driven by piezoelectric actuators, that have fewer moving mass and therefore allow even more injections in a very short period of time. The injection pressure of modern CR systems ranges from 140 MPa to 270 MPa.

Types

There are several different ways of categorising diesel engines, based on different design characteristics given below.

By Power Output

- Small <188 kW (252 hp)

- Medium 188–750 kW

- Large >750 kW

Cylinder Bore

- Passenger car engines: 75 to 100 mm

- Lorry and commercial vehicle engines: 90 to 170 mm

- High-performance high-speed engines: 165 to 280 mm

- Medium-speed engines: 240 to 620 mm

- Low-speed two-stroke engines: 260 to 900 mm

By Number of Strokes

- Four-stroke cycle

- Two-stroke cycle

Piston and Connecting Rod

- Crosshead piston

- Double-acting piston

- Opposed piston

- Trunk piston

By Cylinder Arrangement

Regular cylinder configurations such as straight (inline), V, and boxer (flat) configurations can be used for diesel engines. The inline-six-cylinder design is the most prolific in light- to medium-duty engines, though inline-four engines are also common. Small-capacity engines (generally considered to be those below five litres in capacity) are generally four- or six-cylinder types, with the four-cylinder being the most common type found in automotive uses. The V configuration used to be common for commercial vehicles, but it has been abandoned in favour of the inline configuration.

By Engine Speeds

Günter Mau categorises diesel engines by their rotational speeds into three groups:

- High-speed engines (> 1,000 rpm),

- Medium-speed engines (300–1,000 rpm), and

- Slow-speed engines (< 300 rpm).

High-speed Engines

High-speed engines are used to power trucks (lorries), buses, tractors, cars, yachts, compressors, pumps and small electrical generators. As of 2018, most high-speed engines have direct injection. Many modern engines, particularly in on-highway applications, have common rail direct injection. On bigger ships, high-speed diesel engines are often used for powering electric generators. The highest power output of high-speed diesel engines is approximately 5 MW.

Medium-speed Engines

Medium-speed engines are used in large electrical generators, ship propulsion and mechanical drive applications such as large compressors or pumps. Medium speed diesel engines operate on either diesel fuel or heavy fuel oil by direct injection in the same manner as low-speed engines. Usually, they are four-stroke engines with trunk pistons.

The power output of medium-speed diesel engines can be as high as 21,870 kW, with the effective efficiency being around 47-48%. Most larger medium-speed engines are started with compressed air direct on pistons, using an air distributor, as opposed to a pneumatic starting motor acting on the flywheel, which tends to be used for smaller engines.

Medium-speed engines intended for marine applications are usually used to power (ro-ro) ferries, passenger ships or small freight ships. Using medium-speed engines reduces the cost of smaller ships and increases their transport capacity. In addition to that, a single ship can use two smaller engines instead of one big engine, which increases the ship's safety.

Low-speed Engines

The MAN B&W 5S50MC 5-cylinder, 2-stroke, low-speed marine diesel engine.
This particular engine is found aboard a 29,000 tonne chemical carrier.

Low-speed diesel engines are usually very large in size and mostly used to power ships. There are two different types of low-speed engines that are commonly used: Two-stroke engines with a crosshead, and four-stroke engines with a regular trunk-piston. Two-stroke engines have a limited rotational frequency and their charge exchange is more difficult, which means that they are usually bigger than four-stroke engines and used to directly power a ship's propeller. Four-stroke engines on ships are usually used to power an electric generator. An electric motor powers the propeller. Both types are usually very undersquare. Low-speed diesel engines (as used in ships and other applications where overall engine weight is relatively unimportant) often have an effective efficiency of up to 55%. Like medium-speed engines, low-speed engines are started with compressed air, and they use heavy oil as their primary fuel.

Two-stroke Engines

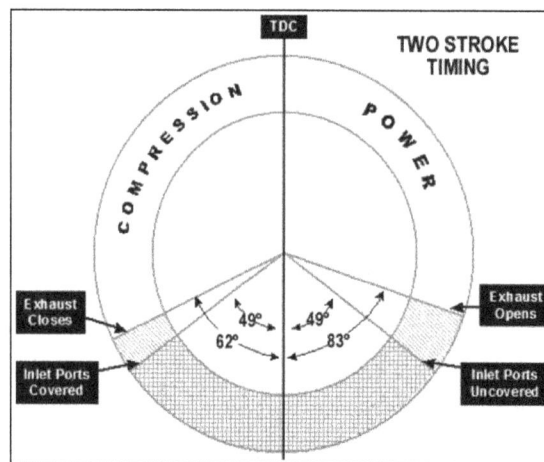

Detroit Diesel timing.

Two-stroke diesel engines use only two strokes instead of four strokes for a complete engine cycle. Filling the cylinder with air and compressing it takes place in one stroke, and the power and exhaust strokes are combined. The compression in a two-stroke diesel engine is similar to the compression that takes place in a four-stroke diesel engine: As the piston passes through bottom centre and starts upward, compression commences, culminating in fuel injection and ignition. Instead of a full set of valves, two-stroke diesel engines have simple intake ports, and exhaust ports (or exhaust valves). When the piston approaches bottom dead centre, both the intake and the exhaust ports are "open", which means that there is atmospheric pressure inside the cylinder. Therefore, some sort of pump is required to blow the air into the cylinder and the combustion gasses into the exhaust. This process is called *scavenging*. The pressure required is approximately 10 to 30 kPa.

Scavenging

In general, there are three types of scavenging possible:

- Uniflow scavenging,

- Crossflow scavenging,

- Reverse flow scavenging.

Crossflow scavenging is incomplete and limits the stroke, yet some manufacturers used it. Reverse flow scavenging is a very simple way of scavenging, and it was popular amongst manufacturers until the early 1980s. Uniflow scavenging is more complicated to make but allows the highest fuel efficiency; since the early 1980s, manufacturers such as MAN and Sulzer have switched to this system. It is standard for modern marine two-stroke diesel engines.

Dual-fuel Diesel Engines

So-called dual-fuel diesel engines or gas diesel engines burn two different types of fuel *simultaneously*, for instance, a gaseous fuel and diesel engine fuel. The diesel engine fuel auto-ignites due to compression ignition, and then ignites the gaseous fuel. Such engines do not require any type of spark ignition and operate similar to regular diesel engines.

Diesel Engine Particularities

Torque and Power

Torque is a force applied to a lever at a right angle multiplied by the lever length. This means that the torque an engine produces depends on the displacement of the engine and the force that the gas pressure inside the cylinder applies to the piston, commonly referred to as *effective piston pressure*:

$$M = p_e \cdot V_h \cdot \pi^{-1} \cdot i^{-1}$$

where M is Torque [N·m]; p_e.. Effective piston pressure [kN·m^{-2}]; V_h.. Displacement [dm³]; i.. Strokes [either 2 or 4].

Example:

- Engine A: effective piston pressure=570 kN·m^{-2}, displacement= 2.2 dm³, strokes= 4, torque= 100 N·m

$$570 \cdot 2.2 \cdot \pi^{-1} \cdot 4^{-1} \approx 100$$

Power is the quotient of work and time:

$$P = 2\pi n M$$

where P is Power [W]; M .. Torque [N·m]; n .. Time (crankshaft speed) [s^{-1}].

which means:

$$P = 2\pi \cdot n_1 \cdot M \cdot 60^{-1}$$

where P is Power [W]; M .. Torque [N·m]; n_1 .. Time (crankshaft speed) [min^{-1}].

Example:

- Engine A: Power≈ 44,000 W, torque= 100 N·m, time= 4200 min^{-1}

$$44,000 \approx 2 \cdot \pi \cdot 4200 \cdot 100 \cdot 60^{-1}$$

- Engine B: Power≈ 44,000 W, torque= 260 N·m, time= 1600 min^{-1}

$$44,000 \approx 2 \cdot \pi \cdot 1600 \cdot 260 \cdot 60^{-1}$$

This means, that increasing either torque or time will result in an increase in power. As the maximum rotational frequency of the diesel engine's crankshaft is usually in between 3500-5000 min^{-1} due to diesel principle limitations, the torque of the diesel engine must be great to achieve a high power, or, in other words, as the diesel engine cannot use a lot of time for achieving a certain amount of power, it has to perform more work and produce more torque.

Mass

The average diesel engine has a poorer power-to-mass ratio than the Otto engine. This is because the diesel must operate at lower engine speeds. Due to the higher operating pressure inside the combustion chamber, which increases the forces on the parts due to inertial forces, the diesel engine needs heavier, stronger parts capable of resisting these forces, which results in an overall greater engine mass.

Emissions

As diesel engines burn a mixture of fuel and air, the exhaust therefore contains substances that consist of the same chemical elements, as fuel and air. The main elements of air are nitrogen (N_2) and oxygen (O_2), fuel consists of hydrogen (H_2) and carbon (C). Burning the fuel will result in the final stage of oxidation. An *ideal diesel engine*, (a hypothetical model that we use as an example), running on an ideal air-fuel mixture, produces an exhaust that consists of carbon dioxide (CO_2), water (H_2O), nitrogen (N_2), and the remaining oxygen (O_2). The combustion process in a real engine differs from an ideal engine's combustion process, and due to incomplete combustion, the exhaust contains additional substances, most notably, carbon monoxide (CO), diesel particulate matter (PM), and due to dissociation, nitrogen oxide (NOx).

When diesel engines burn their fuel with high oxygen levels, this results in high combustion temperatures and higher efficiency, and particulate matter tends to burn, but the amount of NOx pollution tends to increase. NOx pollution can be reduced by recirculating a portion of an engine's exhaust gas back to the engine cylinders, which reduces the oxygen quantity, causing a reduction of combustion temperature, and resulting in fewer NOx. To further reduce NOx emissions, lean NOx traps (LNTs) and SCR-catalysts can be used. Lean NOx traps adsorb the nitrogen oxide and "trap" it. Once the LNT is full, it has to be "regenerated" using hydrocarbons. This is achieved by using a very rich air-fuel mixture, resulting in incomplete combustion. An SCR-catalyst converts nitrogen oxide using urea, which is injected into the exhaust stream, and catalytically converts the NOx into nitrogen (N_2) and water (H_2O). Compared with an Otto engine, the diesel engine produces approximately the same amount of NOx , but some older diesel engines may have an exhaust that contains up to 50% less NOx. However, Otto engines, unlike diesel engines, can use a three-way-catalyst, that converts most of the NOx.

Diesel engine exhaust composition.		
Species	Mass percentage	Volume percentage
Nitrogen (N_2)	75.2%	72.1%
Oxygen (O_2)	15%	0.7%
Carbon dioxide (CO_2)	7.1%	12.3%
Water (H_2O)	2.6%	13.8%
Carbon monoxide (CO)	0.043%	0.09%
Nitrogen oxide (NO_x)	0.034%	0.13%
Hydrocarbons (HC)	0.005%	0.09%
Aldehyde	0.001%	(n/a)
Particulate matter (Sulfate + solid substances)	0.008%	0.0008%

Noise

The distinctive noise of a diesel engine is variably called diesel clatter, diesel nailing, or diesel knock. Diesel clatter is caused largely by the way the fuel ignites; the sudden ignition of the diesel fuel when injected into the combustion chamber causes a pressure wave, resulting in an audible knock. Engine designers can reduce diesel clatter through: Indirect injection; pilot or pre-injection; injection timing; injection rate; compression ratio; turbo boost; and exhaust gas recirculation (EGR). Common rail diesel injection systems permit multiple injection events as an aid to noise reduction. Therefore, newer diesel engines do not knock anymore. Diesel fuels with a higher cetane rating are more likely to ignite and hence reduce diesel clatter.

Cold Weather Starting

In general, diesel engines do not require any starting aid. In cold weather however, some diesel engines can be difficult to start and may need preheating depending on the combustion chamber design. The minimum starting temperature that allows starting without pre-heating is 40 °C for precombustion chamber engines, 20 °C for swirl chamber engines, and 0 °C for direct injected engines. Smaller engines with a displacement of less than 1 litre per cylinder usually have glowplugs, whilst larger heavy-duty engines have flame-start systems.

In the past, a wider variety of cold-start methods were used. Some engines, such as Detroit Diesel engines used a system to introduce small amounts of ether into the inlet manifold to start combustion. Instead of glowplugs, some diesel engines are equipped with starting aid systems that change valve timing. The simplest way this can be done is with a decompression lever. Activating the decompression lever locks the outlet valves in a slight down position, resulting in the engine not having any compression and thus allowing for turning the crankshaft over without resistance. When the crankshaft reaches a higher speed, flipping the decompression lever back into its normal position will abruptly re-activate the outlet valves, resulting in compression – the flywheel's mass moment of inertia then starts the engine. Other diesel engines, such as the precombustion chamber engine XII Jv 170/240 made by Ganz & Co., have a valve timing changing system that is operated by adjusting the inlet valve camshaft, moving it into a slight "late" position. This will make the inlet valves open with a delay, forcing the inlet air to heat up when entering the combustion chamber.

Supercharging and Turbocharging

As the diesel engine relies on manipulation of λ_v for torque controlling and speed regulation, the intake air mass does not have to precisely match the injected fuel mass (which would be $\lambda = 1$). diesel engines are thus ideally suited for supercharging and turbocharging. An additional advantage of the diesel engine is the lack of fuel during the compression stroke. In diesel engines, the fuel is injected near top dead centre (TDC), when the piston is near its highest position. The fuel then ignites due to compression heat. Preignition, caused by the artificial turbocharger compression increase during the compression stroke, cannot occur.

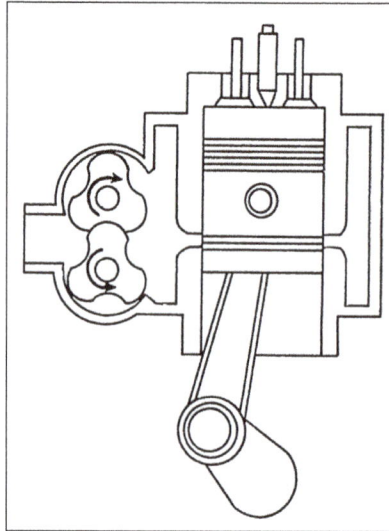

Two stroke diesel engine with Roots blower, typical of Detroit Diesel
and some Electro-Motive Diesel Engines.

Turbocharged 1980s passenger car diesel engine with wastegate
turbocharger and without intercooler (BMW M21).

Many diesels are therefore turbocharged and some are both turbocharged and supercharged. A turbocharged engine can produce more power than a naturally aspirated engine of the same configuration. A supercharger is powered mechanically by the engine's crankshaft, while a turbocharger is powered by the engine exhaust. Turbocharging can improve the fuel economy of diesel engines by recovering waste heat from the exhaust, increasing the excess air factor, and increasing the ratio of engine output to friction losses. Adding an intercooler to a turbocharged engine further increases engine performance by cooling down the air-mass and thus allowing more air-mass per volume.

A two-stroke engine does not have a discrete exhaust and intake stroke and thus is incapable of self-aspiration. Therefore, all two-stroke diesel engines must be fitted with a blower or some form of compressor to charge the cylinders with air and assist in dispersing exhaust gases, a process referred to as scavenging. Roots-type superchargers were used for ship engines until the mid-1950s, since 1955 they have been widely replaced by turbochargers. Usually, a two-stroke ship diesel engine has a single-stage turbocharger with a turbine that has an axial inflow and a radial outflow.

Fuel and Fluid Characteristics

In diesel engines, a mechanical injector system vaporises the fuel directly into the combustion chamber (as opposed to a Venturi jet in a carburetor, or a fuel injector in a manifold injection system vaporising fuel into the intake manifold or intake runners as in a petrol engine). This *forced vaporisation* means that less-volatile fuels can be used. More crucially, because only air is inducted into the cylinder in a diesel engine, the compression ratio can be much higher as there is no risk of pre-ignition provided the injection process is accurately timed. This means that cylinder temperatures are much higher in a diesel engine than a petrol engine, allowing less volatile fuels to be used.

The MAN 630's M-System diesel engine is a petrol engine,
but it also runs on jet fuel, kerosine and diesel engine fuel.

Therefore, diesel engines can operate on a huge variety of different fuels. In general, fuel for diesel engines should have a proper viscosity, so that the injection pump can pump the fuel to the injection nozzles without causing damage to itself or corrosion of the fuel line. At injection, the fuel should form a good fuel spray, and it should not have a coking effect upon the injection nozzles. To ensure proper engine starting and smooth operation, the fuel should be willing to ignite and hence not cause a high ignition delay, (this means that the fuel should have a high cetane number). Diesel fuel should also have a high lower heating value.

Inline mechanical injector pumps generally tolerate poor-quality or bio-fuels better than distributor-type pumps. Also, indirect injection engines generally run more satisfactorily on fuels with a high ignition delay (for instance, petrol) than direct injection engines. This is partly because an indirect injection engine has a much greater 'swirl' effect, improving vaporisation and combustion of fuel, and because (in the case of vegetable oil-type fuels) lipid depositions can condense on the cylinder walls of a direct-injection engine if combustion temperatures are too low (such as starting the engine from cold). Direct-injected engines with an MAN centre sphere combustion chamber rely on fuel condensing on the combustion chamber walls. The fuel starts vaporising only after ignition sets in, and it burns relatively smoothly. Therefore, such engines also tolerate fuels with poor ignition delay characteristics, and, in general, they can operate on petrol rated 86 RON.

Fuel Types

In his 1893 work *Theory and Construction of a Rational Heat Motor*, Rudolf Diesel considers using coal dust as fuel for the diesel engine. However, Diesel just *considered* using coal dust (as well as liquid fuels and gas); his actual engine was designed to operate on petroleum, which was soon replaced with regular petrol and kerosine for further testing purposes, as petroleum proved to be too viscous. In addition to kerosine and petrol, Diesel's engine could also operate on ligroin.

Before diesel engine fuel was standardised, fuels such as petrol, kerosine, gas oil, vegetable oil and mineral oil, as well as mixtures of these fuels, were used. Typical fuels specifically intended to be used for diesel engines were petroleum distillates and coal-tar distillates such as the following; these fuels have specific lower heating values of:

- Diesel oil: 10,200 kcal·kg^{-1} (42.7 MJ·kg^{-1}) up to 10,250 kcal·kg^{-1} (42.9 MJ·kg^{-1}).

- Heating oil: 10,000 kcal·kg^{-1} (41.8 MJ·kg^{-1}) up to 10,200 kcal·kg^{-1} (42.7 MJ·kg^{-1}).

- Coal-tar creosote: 9,150 kcal·kg^{-1} (38.3 MJ·kg^{-1}) up to 9,250 kcal·kg^{-1} (38.7 MJ·kg^{-1}).

- Kerosine: up to 10,400 kcal·kg^{-1} (43.5 MJ·kg^{-1}).

The first diesel fuel standards were the DIN 51601, VTL 9140-001, and NATO F 54, which appeared after World War II. The modern European EN 590 diesel fuel standard was established in May 1993; the modern version of the NATO F 54 standard is mostly identical with it. The DIN 51628 biodiesel standard was rendered obsolete by the 2009 version of the EN 590; FAME biodiesel conforms to the EN 14214 standard. Watercraft diesel engines usually operate on diesel engine fuel that conforms to the ISO 8217 standard (Bunker C). Also, some diesel engines can operate on gasses (such as LNG).

Modern Diesel Fuel Properties

Modern diesel fuel properties		
	EN 590 (as of 2009)	EN 14214 (as of 2010)
Ignition performance	≥ 51 CN	≥ 51 CN
Density at 15 °C	820 to 845 kg·m^{-3}	860 to 900 kg·m^{-3}
Sulphur content	≤10 mg·kg^{-1}	≤10 mg·kg^{-1}
Water content	≤200 mg·kg^{-1}	≤500 mg·kg^{-1}
Lubricity	460 μm	460 μm
Viscosity at 40 °C	2.0 to 4.5 mm^2·s^{-1}	3.5 to 5.0 mm^2·s^{-1}
FAME content	≤7.0%	≥96.5%
Molar H/C ratio	–	1.69
Lower heating value	–	37.1 MJ·kg^{-1}

Gelling

DIN 51601 diesel fuel was prone to waxing or gelling in cold weather; both are terms for the solidification of diesel oil into a partially crystalline state. The crystals build up in the fuel system (especially in fuel filters), eventually starving the engine of fuel and causing it to stop running.

Low-output electric heaters in fuel tanks and around fuel lines were used to solve this problem. Also, most engines have a spill return system, by which any excess fuel from the injector pump and injectors is returned to the fuel tank. Once the engine has warmed, returning warm fuel prevents waxing in the tank. Some manufacturers, such as BMW, recommended fuelling diesel cars with petrol to prevent the fuel from gelling when the temperatures dropped below −15 °C.

Safety

Fuel Flammability

Diesel fuel is less flammable than petrol, because its flash point is 55 °C, leading to a lower risk of fire caused by fuel in a vehicle equipped with a diesel engine.

Diesel fuel can create an explosive air/vapour mix under the right conditions. However, compared with petrol, it is less prone due to its lower vapour pressure, which is an indication of evaporation rate. The Material Safety Data Sheet for ultra-low sulfur diesel fuel indicates a vapour explosion hazard for diesel fuel indoors, outdoors, or in sewers.

Cancer

Diesel exhaust has been classified as an IARC Group 1 carcinogen. It causes lung cancer and is associated with an increased risk for bladder cancer.

Applications

The characteristics of diesel have different advantages for different applications.

Passenger Cars

Diesel engines have long been popular in bigger cars and have been used in smaller cars such as superminis in Europe since the 1980s. They were popular in larger cars earlier, as the weight and cost penalties were less noticeable. Smooth operation as well as high low end torque are deemed important for passenger cars and small commercial vehicles. The introduction of electronically controlled fuel injection significantly improved the smooth torque generation, and starting in the early 1990s, car manufacturers began offering their high-end luxury vehicles with diesel engines. Passenger car diesel engines usually have between three and ten cylinders, and a displacement ranging from 0.8 to 5.0 litres. Modern powerplants are usually turbocharged and have direct injection.

Diesel engines do not suffer from intake-air throttling, resulting in very low fuel consumption especially at low partial load (for instance: driving at city speeds). One fifth of all passenger cars worldwide have diesel engines, with many of them being in Europe, where approximately 47% of all passenger cars are diesel-powered. Daimler-Benz in conjunction with Robert Bosch GmbH produced diesel-powered passenger cars starting in 1936. The popularity of diesel-powered passenger cars in markets such as India, South Korea and Japan is increasing (as of 2018).

Commercial Vehicles and Lorries

In 1893, Rudolf Diesel suggested that the diesel engine could possibly power 'wagons' (lorries). The first lorries with diesel engines were brought to market in 1924.

Modern diesel engines for lorries have to be both extremely reliable and very fuel efficient. Common-rail direct injection, turbocharging and four valves per cylinder are standard. Displacements range from 4.5 to 15.5 litres, with power-to-mass ratios of 2.5–3.5 kg·kW^{-1} for heavy duty and 2.0–3.0 kg·kW^{-1} for medium duty engines. V6 and V8 engines used to be common, due to the relatively low engine mass the V configuration provides. Recently, the V configuration has been abandoned in favour of straight engines. These engines are usually straight-6 for heavy and medium duties and straight-4 for medium duty. Their undersquare design causes lower overall piston speeds which results in increased lifespan of up to 1,200,000 km. Compared with 1970s diesel engines, the expected lifespan of modern lorry diesel engines has more than doubled.

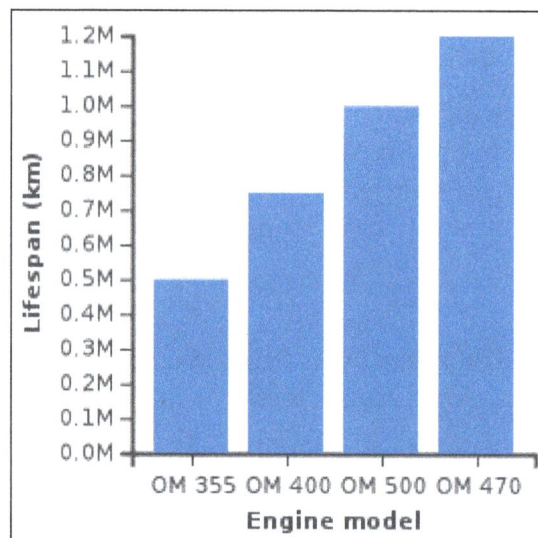

Lifespan of Mercedes-Benz diesel engines.

Railroad Rolling Stock

Diesel engines for locomotives are built for continuous operation and may require the ability to use poor quality fuel in some circumstances. Some locomotives use two-stroke diesel engines. Diesel engines have eclipsed steam engines as the prime mover on all non-electrified railroads in the industrialised world. The first diesel locomotives appeared in 1913, and diesel multiple units soon after. Many modern diesel locomotives are actually diesel-electric locomotives: the diesel engine is used to power an electric generator that in turn powers electric traction motors with no mechanical connection between diesel engine and traction. While electric locomotives have replaced the diesel locomotive for some passenger traffic in Europe and Asia, diesel is still today very popular for cargo-hauling freight trains and on tracks where electrification is not feasible.

In the 1940s, road vehicle diesel engines with power outputs of 150 to 200 PS (110 to 147 kW) were considered reasonable for DMUs. Commonly, regular truck powerplants were used. The height of these engines had to be less than 1,000 mm to allow underfloor installation. Usually, the engine was mated with a pneumatically operated mechanical gearbox, due to the low size, mass, and production costs of this design. Some DMUs used hydraulic torque converters instead. Diesel-electric transmission was not suitable for such small engines. In the 1930s, the Deutsche Reichsbahn standardised its first DMU engine. It was a 30.3 litre, 12-cylinder boxer unit, producing 275 PS (202 kW). Several German manufacturers produced engines according to this standard.

Watercraft

The requirements for marine diesel engines vary, depending on the application. For military use and medium-size boats, medium-speed four-stroke diesel engines are most suitable. These engines usually have up to 24 cylinders and come with power outputs in the one-digit Megawatt region. Small boats may use lorry diesel engines. Large ships use extremely efficient, low-speed two-stroke diesel engines. They can reach efficiencies of up to 55%. Unlike most regular diesel engines, two-stroke watercraft engines use highly viscous fuel oil. Submarines are usually diesel-electric.

One of the eight-cylinder 3200 I.H.P. Harland and Wolff – Burmeister & Wain diesel engines installed in the motorship *Glenapp*. This was the highest powered diesel engine yet installed in a ship. Note man standing lower right for size comparison.

The first diesel engines for ships were made by A. B. Diesels Motorer Stockholm in 1903. These engines were three-cylinder units of 120 PS (88 kW) and four-cylinder units of 180 PS (132 kW) and used for Russian ships. In World War I, especially submarine diesel engine development advanced quickly. By the end of the War, double acting piston two-stroke engines with up to 12,200 PS (9 MW) had been made for marine use.

Non-road Diesel Engines

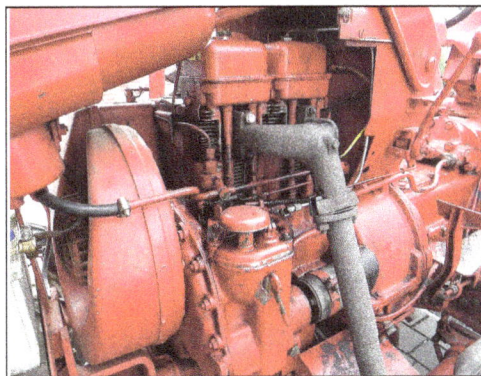

Air-cooled diesel engine of a 1959 Porsche 218.

Non-road diesel engines are commonly used for construction equipment. Fuel efficiency, reliability and ease of maintenance are very important for such engines, whilst high power output and quiet operation are negligible. Therefore, mechanically controlled fuel injection and air-cooling are still very common. The common power outputs of non-road diesel engines vary a lot, with the smallest units starting at 3 kW, and the most powerful engines being heavy duty lorry engines.

Stationary Diesel Engines

Stationary diesel engines are commonly used for electricity generation, but also for powering refrigerator compressors, or other types of compressors or pumps. Usually, these engines run permanently, either with mostly partial load, or intermittently, with full load. Stationary diesel engines powering electric generators that put out an alternating current, usually operate with alternating load, but fixed rotational frequency. This is due to the mains' fixed frequency of either 50 Hz (Europe), or 60 Hz (United States). The engine's crankshaft rotational frequency is chosen so that the mains' frequency is a multiple of it. For practical reasons, this results in crankshaft rotational frequencies of either 25 Hz (1500 per minute) or 30 Hz (1800 per minute).

English Electric.

Low Heat Rejection Engines

A special class of prototype internal combustion piston engines has been developed over several decades with the goal of improving efficiency by reducing heat loss. These engines are variously called adiabatic engines; due to better approximation of adiabatic expansion; low heat rejection engines, or high temperature engines. They are generally piston engines with combustion chamber parts lined with ceramic thermal barrier coatings. Some make use of pistons and other parts made of titanium which has a low thermal conductivity and density. Some designs are able to eliminate the use of a cooling system and associated parasitic losses altogether. Developing lubricants able to withstand the higher temperatures involved has been a major barrier to commercialization.

Steam Engines

A model of a beam engine featuring James Watt's parallel linkage for double action.

A mill engine from Stott Park Bobbin Mill, Cumbria, England.

A steam locomotive from East Germany.

A steam engine is a heat engine that performs mechanical work using steam as its working fluid. The steam engine uses the force produced by steam pressure to push a piston back and forth inside a cylinder. This pushing force is transformed, by a connecting rod and flywheel, into rotational force for work. The term "steam engine" is generally applied only to reciprocating engines as just described, not to the steam turbine.

Steam engines are external combustion engines, where the working fluid is separated from the combustion products. The ideal thermodynamic cycle used to analyze this process is called the Rankine cycle.

In general usage, the term *steam engine* can refer to either complete steam plants (including boilers etc.) such as railway steam locomotives and portable engines, or may refer to the piston or turbine machinery alone, as in the beam engine and stationary steam engine.

Steam-driven devices were known as early as the aeolipile in the first century AD, with a few other uses recorded in the 16th and 17th century. Thomas Savery's dewatering pump used steam pressure operating directly on the water. The first commercially successful engine that could transmit continuous power to a machine was developed in 1712 by Thomas Newcomen. James Watt made a critical improvement by removing spent steam to a separate vessel for condensation, greatly improving the amount of work obtained per unit of fuel consumed. By the 19th century, stationary steam engines powered the factories of the Industrial Revolution. Steam engines replaced sail for ships, and steam locomotives operated on the railways.

Reciprocating piston type steam engines were the dominant source of power until the early 20th century, when advances in the design of electric motors and internal combustion engines gradually resulted in the replacement of reciprocating (piston) steam engines in commercial usage. Steam turbines replaced reciprocating engines in power generation, due to lower cost, higher operating speed, and higher efficiency.

Components and Accessories of Steam Engines

There are two fundamental components of a steam plant: the boiler or steam generator, and the "motor unit", referred to itself as a "steam engine". Stationary steam engines in fixed buildings may have the boiler and engine in separate buildings some distance apart. For portable or mobile use, such as steam locomotives, the two are mounted together.

The widely used reciprocating engine typically consisted of a cast-iron cylinder, piston, connecting rod and beam or a crank and flywheel, and miscellaneous linkages. Steam was alternately supplied and exhausted by one or more valves. Speed control was either automatic, using a governor, or by a manual valve. The cylinder casting contained steam supply and exhaust ports.

Engines equipped with a condenser are a separate type than those that exhaust to the atmosphere.

Other components are often present; pumps (such as an injector) to supply water to the boiler during operation, condensers to recirculate the water and recover the latent heat of vaporisation, and superheaters to raise the temperature of the steam above its saturated vapour point, and various mechanisms to increase the draft for fireboxes. When coal is used, a chain or screw stoking mechanism and its drive engine or motor may be included to move the fuel from a supply bin (bunker) to the firebox.

Heat Source

The heat required for boiling the water and raising the temperature of the steam can be derived from various sources, most commonly from burning combustible materials with an appropriate supply of air in a closed space (called variously combustion chamber, firebox, furnace). In the case of model or toy steam engines, the heat source can be an electric heating element.

Boilers

An industrial boiler used for a stationary steam engine.

Boilers are pressure vessels that contain water to be boiled, and features that transfer the heat to the water as effectively as possible.

The two most common types are:

1. Water-tube boiler – Water is passed through tubes surrounded by hot gas.

2. Fire-tube boiler – Hot gas is passed through tubes immersed in water, the same water also circulates in a water jacket surrounding the firebox and, in high-output locomotive boilers, also passes through tubes in the firebox itself (thermic syphons and security circulators).

Fire-tube boilers were the main type used for early high-pressure steam (typical steam locomotive practice), but they were to a large extent displaced by more economical water tube boilers in the late 19th century for marine propulsion and large stationary applications.

Many boilers raise the temperature of the steam after it has left that part of the boiler where it is in contact with the water. Known as superheating it turns 'wet steam' into 'superheated steam'. It avoids the steam condensing in the engine cylinders, and gives a significantly higher efficiency.

Motor Units

In a steam engine, a piston or steam turbine or any other similar device for doing mechanical work takes a supply of steam at high pressure and temperature and gives out a supply of steam at lower pressure and temperature, using as much of the difference in steam energy as possible to do mechanical work.

These "motor units" are often called 'steam engines' in their own right. Engines using compressed air or other gases differ from steam engines only in details that depend on the nature of the gas although compressed air has been used in steam engines without change.

Cold Sink

As with all heat engines, the majority of primary energy must be emitted as waste heat at relatively low temperature.

The simplest cold sink is to vent the steam to the environment. This is often used on steam locomotives to avoid the weight and bulk of condensers. Some of the released steam is vented up the chimney so as to increase the draw on the fire, which greatly increases engine power, but reduces efficiency.

Sometimes the waste heat from the engine is useful itself, and in those cases, very high overall efficiency can be obtained.

Steam engines in stationary power plants use surface condensers as a cold sink. The condensers are cooled by water flow from oceans, rivers, lakes, and often by cooling towers which evaporate water to provide cooling energy removal. The resulting condensed hot water (*condensate*), is then pumped back up to pressure and sent back to the boiler. A dry-type cooling tower is similar to an automobile radiator and is used in locations where water is costly. Waste heat can also be ejected by evaporative (wet) cooling towers, which use a secondary external water circuit that evaporates some of flow to the air.

River boats initially used a jet condenser in which cold water from the river is injected into the exhaust steam from the engine. Cooling water and condensate mix. While this was also applied for sea-going vessels, generally after only a few days of operation the boiler would become coated with deposited salt, reducing performance and increasing the risk of a boiler explosion. Starting about 1834, the use of surface condensers on ships eliminated fouling of the boilers, and improved engine efficiency.

Evaporated water cannot be used for subsequent purposes (other than rain somewhere), whereas river water can be re-used. In all cases, the steam plant boiler feed water, which must be kept pure, is kept separate from the cooling water or air.

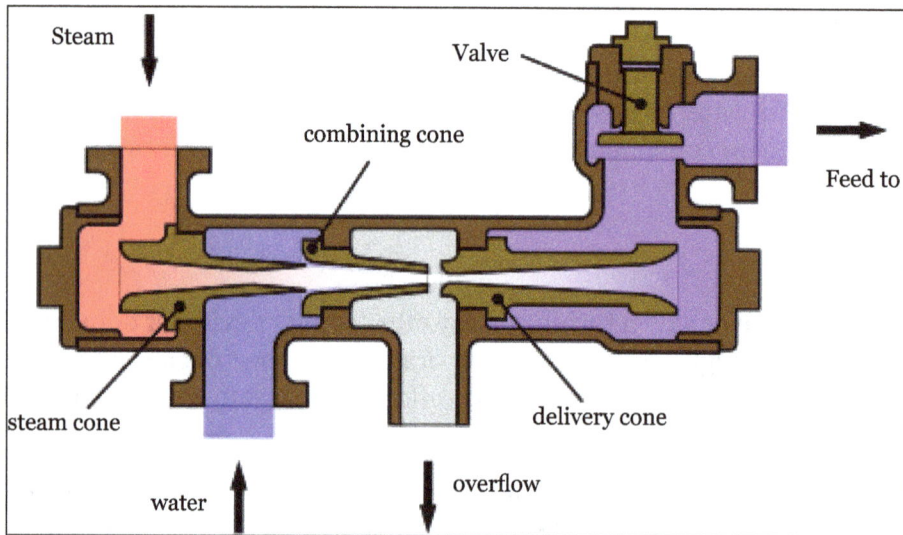

An injector uses a jet of steam to force water into the boiler. Injectors are inefficient but simple enough to be suitable for use on locomotives.

Water Pump

The Rankine cycle and most practical steam engines have a water pump to recycle or top up the boiler water, so that they may be run continuously. Utility and industrial boilers commonly use multi-stage centrifugal pumps; however, other types are used. Another means of supplying lower-pressure boiler feed water is an injector, which uses a steam jet usually supplied from the boiler. Injectors became popular in the 1850s but are no longer widely used, except in applications such as steam locomotives. It is the pressurization of the water that circulates through the steam boiler that allows the water to be raised to temperatures well above 100 °C (212 °F) boiling point of water at one atmospheric pressure, and by that means to increase the efficiency of the steam cycle.

Monitoring and Control

For safety reasons, nearly all steam engines are equipped with mechanisms to monitor the boiler, such as a pressure gauge and a sight glass to monitor the water level.

Many engines, stationary and mobile, are also fitted with a governor to regulate the speed of the engine without the need for human interference.

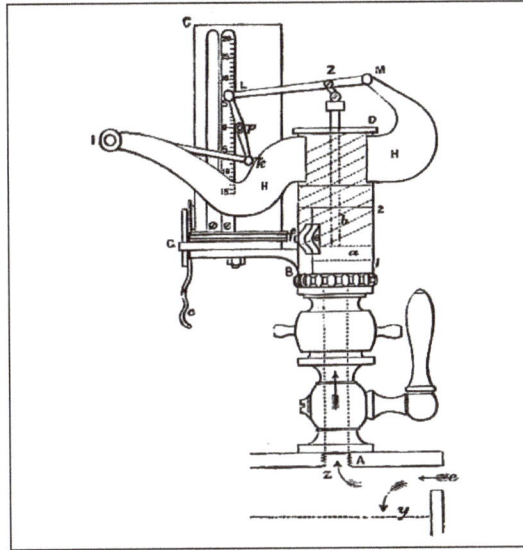

Richard's indicator instrument of 1875.

The most useful instrument for analyzing the performance of steam engines is the steam engine indicator. Early versions were in use by 1851, but the most successful indicator was developed for the high speed engine inventor and manufacturer Charles Porter by Charles Richard and exhibited at London Exhibition in 1862. The steam engine indicator traces on paper the pressure in the cylinder throughout the cycle, which can be used to spot various problems and calculate developed horsepower. It was routinely used by engineers, mechanics and insurance inspectors. The engine indicator can also be used on internal combustion engines.

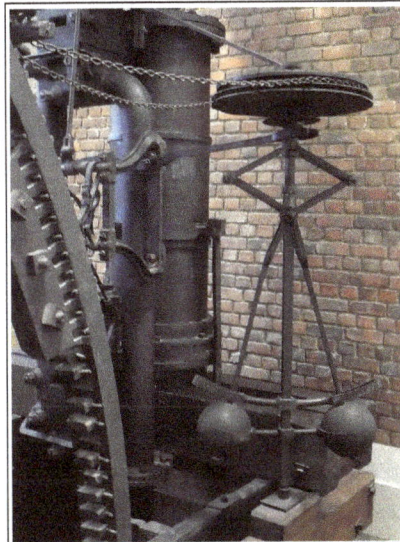

Centrifugal governor in the Boulton & Watt engine.

Governor

The centrifugal governor was adopted by James Watt for use on a steam engine in 1788 after Watt's partner Boulton saw one on the equipment of a flour mill Boulton & Watt were building. The governor could not actually hold a set speed, because it would assume a new constant speed in response to load changes. The governor was able to handle smaller variations such as those caused by fluctuating heat load to the boiler. Also, there was a tendency for oscillation whenever there was a speed change. As a consequence, engines equipped only with this governor were not suitable for operations requiring constant speed, such as cotton spinning. The governor was improved over time and coupled with variable steam cut off, good speed control in response to changes in load was attainable near the end of the 19th century.

Engine Configuration

Simple Engine

In a simple engine, or "single expansion engine" the charge of steam passes through the entire expansion process in an individual cylinder, although a simple engine may have one or more individual cylinders. It is then exhausted directly into the atmosphere or into a condenser. As steam expands in passing through a high-pressure engine, its temperature drops because no heat is being added to the system; this is known as adiabatic expansion and results in steam entering the cylinder at high temperature and leaving at lower temperature. This causes a cycle of heating and cooling of the cylinder with every stroke, which is a source of inefficiency.

The dominant efficiency loss in reciprocating steam engines is cylinder condensation and re-evaporation. The steam cylinder and adjacent metal parts/ports operate at a temperature about halfway between the steam admission saturation temperature and the saturation temperature corresponding to the exhaust pressure. As high-pressure steam is admitted into the working cylinder, much of the high-temperature steam is condensed as water droplets onto the metal surfaces, significantly reducing the steam available for expansive work. When the expanding steam reaches low pressure (especially during the exhaust stroke), the previously deposited water droplets that had just been formed within the cylinder/ports now boil away (re-evaporation) and this steam does no further work in the cylinder.

There are practical limits on the expansion ratio of a steam engine cylinder, as increasing cylinder surface area tends to exacerbate the cylinder condensation and re-evaporation issues. This negates the theoretical advantages associated with a high ratio of expansion in an individual cylinder.

Compound Engines

A method to lessen the magnitude of energy loss to a very long cylinder was invented in 1804 by British engineer Arthur Woolf, who patented his Woolf high-pressure compound engine in 1805. In the compound engine, high-pressure steam from the boiler expands in a high-pressure (HP) cylinder and then enters one or more subsequent lower-pressure (LP) cylinders. The complete expansion of the steam now occurs across multiple cylinders, with the overall temperature drop within each cylinder reduced considerably. By expanding the steam in steps with smaller temperature range (within each cylinder) the condensation and re-evaporation efficiency issue is reduced. This reduces the magnitude of cylinder heating and cooling, increasing the efficiency of the engine. By staging the expansion in multiple cylinders, variations of torque can be reduced. To derive equal work from lower-pressure cylinder requires a larger cylinder volume as this steam occupies a greater volume. Therefore, the bore, and in rare cases the stroke, are increased in low-pressure cylinders, resulting in larger cylinders.

Double-expansion (usually known as compound) engines expanded the steam in two stages. The pairs may be duplicated or the work of the large low-pressure cylinder can be split with one high-pressure cylinder exhausting into one or the other, giving a three-cylinder layout where cylinder and piston diameter are about the same, making the reciprocating masses easier to balance.

Two-cylinder compounds can be arranged as:

- Cross compounds: The cylinders are side by side.

- Tandem compounds: The cylinders are end to end, driving a common connecting rod

- Angle compounds: The cylinders are arranged in a V (usually at a 90° angle) and drive a common crank.

With two-cylinder compounds used in railway work, the pistons are connected to the cranks as with a two-cylinder simple at 90° out of phase with each other (*quartered*). When the double-expansion group is duplicated, producing a four-cylinder compound, the individual pistons within the group are usually balanced at 180°, the groups being set at 90° to each other. In one case (the first type of Vauclain compound), the pistons worked in the same phase driving a common crosshead and crank, again set at 90° as for a two-cylinder engine. With the three-cylinder compound arrangement, the LP cranks were either set at 90° with the HP one at 135° to the other two, or in some cases, all three cranks were set at 120°.

The adoption of compounding was common for industrial units, for road engines and almost universal for marine engines after 1880; it was not universally popular in railway locomotives where it was often perceived as complicated. This is partly due to the harsh railway operating environment and limited space afforded by the loading gauge (particularly in Britain, where compounding was never common and not employed after 1930). However, although never in the majority, it was popular in many other countries.

Multiple-expansion Engines

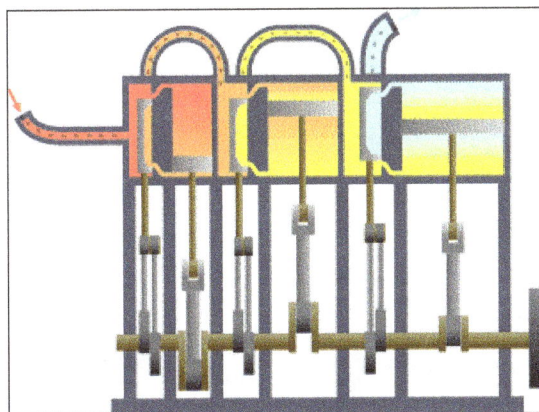

An animation of a simplified triple-expansion engine. High-pressure steam (red) enters from the boiler and passes through the engine, exhausting as low-pressure steam (blue), usually to a condenser.

It is a logical extension of the compound engine (described above) to split the expansion into yet more stages to increase efficiency. The result is the multiple-expansion engine. Such engines use either three or four expansion stages and are known as *triple-* and *quadruple-expansion engines* respectively. These engines use a series of cylinders of progressively increasing diameter. These cylinders are designed to divide the work into equal shares for each expansion stage. As with the double-expansion engine, if space is at a premium, then two smaller cylinders may be used for the low-pressure stage. Multiple-expansion engines typically had the cylinders arranged inline, but various other formations were used. In the late 19th century, the Yarrow-Schlick-Tweedy balancing "system" was used on some marine triple-expansion engines. Y-S-T engines divided the low-pressure expansion stages between two cylinders, one at each end of the engine. This allowed the crankshaft to be better balanced, resulting in a smoother, faster-responding engine which ran with less vibration. This made the four-cylinder triple-expansion engine popular with large passenger liners (such as the *Olympic* class), but this was ultimately replaced by the virtually vibration-free turbine engine. It is noted, however, that triple-expansion reciprocating steam engines were used to drive the World War II Liberty ships, by far the largest number of identical ships ever built. Over 2700 ships were built, in the United States, from a British original design.

The image shows an animation of a triple-expansion engine. The steam travels through the engine from left to right. The valve chest for each of the cylinders is to the left of the corresponding cylinder.

Land-based steam engines could exhaust their steam to atmosphere, as feed water was usually readily available. Prior to and during World War I, the expansion engine dominated marine applications, where high vessel speed was not essential. It was, however, superseded by the British invention steam turbine where speed was required, for instance in warships, such as the dreadnought battleships, and ocean liners. HMS *Dreadnought* of 1905 was the first major warship to replace the proven technology of the reciprocating engine with the then-novel steam turbine.

Types of Motor Units

Reciprocating Piston

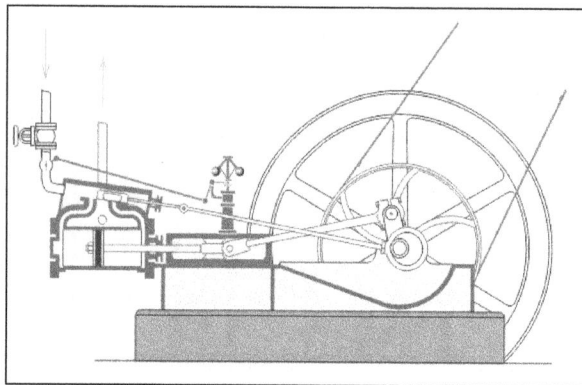

Double acting stationary engine. This was the common mill engine of the mid 19th century.
Note the slide valve with concave, almost "D" shaped, underside.

In most reciprocating piston engines, the steam reverses its direction of flow at each stroke (counterflow), entering and exhausting from the same end of the cylinder. The complete engine cycle occupies one rotation of the crank and two piston strokes; the cycle also comprises four *events* aadmission, expansion, exhaust, compression. These events are controlled by valves often working inside a *steam chest* adjacent to the cylinder; the valves distribute the steam by opening and closing steam *ports* communicating with the cylinder end(s) and are driven by valve gear, of which there are many types.

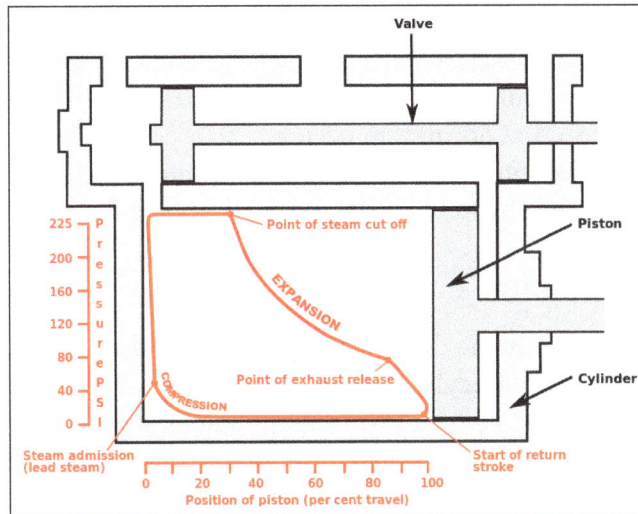

Schematic Indicator diagram showing the four events in a double piston stroke.

The simplest valve gears give events of fixed length during the engine cycle and often make the engine rotate in only one direction. Many however have a reversing mechanism which additionally can provide means for saving steam as speed and momentum are gained by gradually "shortening the cutoff" or rather, shortening the admission event; this in turn proportionately lengthens the expansion period. However, as one and the same valve usually controls both steam flows, a short cutoff at admission adversely affects the exhaust and compression periods which should ideally always be kept fairly constant; if the exhaust event is too brief, the totality of the exhaust steam cannot evacuate the cylinder, choking it and giving excessive compression ("*kick back*").

In the 1840s and 1850s, there were attempts to overcome this problem by means of various patent valve gears with a separate, variable cutoff expansion valve riding on the back of the main slide valve; the latter usually had fixed or limited cutoff. The combined setup gave a fair approximation of the ideal events, at the expense of increased friction and wear, and the mechanism tended to be complicated. The usual compromise solution has been to provide *lap* by lengthening rubbing surfaces of the valve in such a way as to overlap the port on the admission side, with the effect that the exhaust side remains open for a longer period after cut-off on the admission side has occurred. This expedient has since been generally considered satisfactory for most purposes and makes possible the use of the simpler Stephenson, Joy and Walschaerts motions. Corliss, and later, poppet valve gears had separate admission and exhaust valves driven by trip mechanisms or cams profiled so as to give ideal events; most of these gears never succeeded outside of the stationary marketplace due to various other issues including leakage and more delicate mechanisms.

Compression

Before the exhaust phase is quite complete, the exhaust side of the valve closes, shutting a portion of the exhaust steam inside the cylinder. This determines the compression phase where a cushion of steam is formed against which the piston does work whilst its velocity is rapidly decreasing; it moreover obviates the pressure and temperature shock, which would otherwise be caused by the sudden admission of the high-pressure steam at the beginning of the following cycle.

Lead

The above effects are further enhanced by providing *lead*: as was later discovered with the internal combustion engine, it has been found advantageous since the late 1830s to advance the admission phase, giving the valve *lead* so that admission occurs a little before the end of the exhaust stroke in order to fill the *clearance volume* comprising the ports and the cylinder ends (not part of the piston-swept volume) before the steam begins to exert effort on the piston.

Uniflow (or Unaflow) Engine

Uniflow engines attempt to remedy the difficulties arising from the usual counterflow cycle where, during each stroke, the port and the cylinder walls will be cooled by the passing exhaust steam, whilst the hotter incoming admission steam will waste some of its energy in restoring the working temperature. The aim of the uniflow is to remedy this defect and improve efficiency by providing an additional port uncovered by the piston at the end of each stroke making the steam flow only in one direction. By this means, the simple-expansion uniflow engine gives efficiency equivalent to that of classic compound systems with the added advantage of superior part-load performance, and comparable efficiency to turbines for smaller engines below one thousand horsepower. However, the thermal expansion gradient uniflow engines produce along the cylinder wall gives practical difficulties.

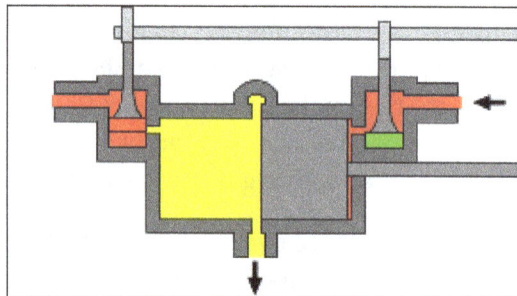

Schematic animation of a uniflow steam engine. The poppet valves are controlled
by the rotating camshaft at the top. High-pressure steam enters, red, and exhausts, yellow.

Turbine Engines

A rotor of a modern steam turbine, used in a power plant.

A steam turbine consists of one or more *rotors* (rotating discs) mounted on a drive shaft, alternating with a series of *stators* (static discs) fixed to the turbine casing. The rotors have a propeller-like arrangement of blades at the outer edge. Steam acts upon these blades, producing rotary motion. The stator consists of a similar, but fixed, series of blades that serve to redirect the steam flow onto the next rotor stage. A steam turbine often exhausts into a surface condenser that provides a vacuum. The stages of a steam turbine are typically arranged to extract the maximum potential work from a specific velocity and pressure of steam, giving rise to a series of variably sized high- and low-pressure stages. Turbines are only efficient if they rotate at relatively high speed, therefore they are usually connected to reduction gearing to drive lower speed applications, such as a ship's propeller. In the vast majority of large electric generating stations, turbines are directly connected to generators with no reduction gearing. Typical speeds are 3600 revolutions per minute (RPM) in the United States with 60 Hertz power, and 3000 RPM in Europe and other countries with 50 Hertz electric power systems. In nuclear power applications, the turbines typically run at half these speeds, 1800 RPM and 1500 RPM. A turbine rotor is also only capable of providing power when rotating in one direction. Therefore, a reversing stage or gearbox is usually required where power is required in the opposite direction.

Steam turbines provide direct rotational force and therefore do not require a linkage mechanism to convert reciprocating to rotary motion. Thus, they produce smoother rotational forces on the output shaft. This contributes to a lower maintenance requirement and less wear on the machinery they power than a comparable reciprocating engine.

Turbinia – the first steam turbine-powered ship.

The main use for steam turbines is in electricity generation (in the 1990s about 90% of the world's electric production was by use of steam turbines) however the recent widespread application of large gas turbine units and typical combined cycle power plants has resulted in reduction of this percentage to the 80% regime for steam turbines. In electricity production, the high speed of turbine rotation matches well with the speed of modern electric generators, which are typically direct connected to their driving turbines. In marine service, (pioneered on the *Turbinia*), steam turbines with reduction gearing (although the Turbinia has direct turbines to propellers with no reduction gearbox) dominated large ship propulsion throughout the late 20th century, being more efficient (and requiring far less maintenance) than reciprocating steam engines. In recent decades, reciprocating Diesel engines, and gas turbines, have almost entirely supplanted steam propulsion for marine applications.

Virtually all nuclear power plants generate electricity by heating water to provide steam that drives a turbine connected to an electrical generator. Nuclear-powered ships and submarines either use a steam turbine directly for main propulsion, with generators providing auxiliary power, or else employ turbo-electric transmission, where the steam drives a turbo generator set with propulsion provided by electric motors. A limited number of steam turbine railroad locomotives were manufactured. Some non-condensing direct-drive locomotives did meet with some success for long haul freight operations in Sweden and for express passenger work in Britain, but were not repeated. Elsewhere, notably in the United States, more advanced designs with electric transmission were built experimentally, but not reproduced. It was found that steam turbines were not ideally suited to the railroad environment and these locomotives failed to oust the classic reciprocating steam unit in the way that modern diesel and electric traction has done.

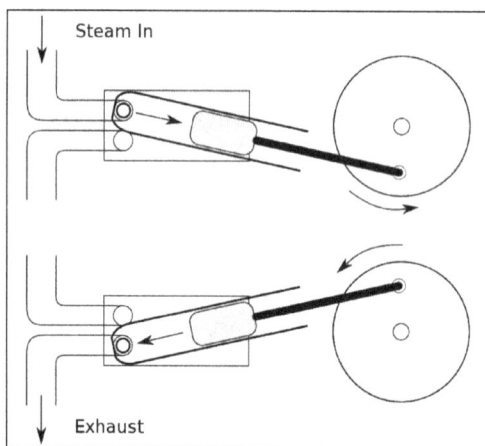

Operation of a simple oscillating cylinder steam engine.

Oscillating Cylinder Steam Engines

An oscillating cylinder steam engine is a variant of the simple expansion steam engine which does not require valves to direct steam into and out of the cylinder. Instead of valves, the entire cylinder rocks, or oscillates, such that one or more holes in the cylinder line up with holes in a fixed port face or in the pivot mounting (trunnion). These engines are mainly used in toys and models, because of their simplicity, but have also been used in full-size working engines, mainly on ships where their compactness is valued.

Rotary Steam Engines

It is possible to use a mechanism based on a pistonless rotary engine such as the Wankel engine in place of the cylinders and valve gear of a conventional reciprocating steam engine. Many such engines have been designed, from the time of James Watt to the present day, but relatively few were actually built and even fewer went into quantity production. The major problem is the difficulty of sealing the rotors to make them steam-tight in the face of wear and thermal expansion; the resulting leakage made them very inefficient. Lack of expansive working, or any means of control of the cutoff, is also a serious problem with many such designs.

By the 1840s, it was clear that the concept had inherent problems and rotary engines were treated with some derision in the technical press. However, the arrival of electricity on the scene, and the obvious advantages of driving a dynamo directly from a high-speed engine, led to something of a revival in interest in the 1880s and 1890s, and a few designs had some limited success.

An aeolipile rotates due to the steam escaping from the arms.
No practical use was made of this effect.

Of the few designs that were manufactured in quantity, those of the Hult Brothers Rotary Steam Engine Company of Stockholm, Sweden, and the spherical engine of Beauchamp Tower are notable. Tower's engines were used by the Great Eastern Railway to drive lighting dynamos on their locomotives, and by the Admiralty for driving dynamos on board the ships of the Royal Navy. They were eventually replaced in these niche applications by steam turbines.

Rocket Type

The aeolipile represents the use of steam by the rocket-reaction principle, although not for direct propulsion.

In more modern times there has been limited use of steam for rocketry – particularly for rocket cars. Steam rocketry works by filling a pressure vessel with hot water at high pressure and opening a valve leading to a suitable nozzle. The drop in pressure immediately boils some of the water and the steam leaves through a nozzle, creating a propulsive force.

Safety

Steam engines possess boilers and other components that are pressure vessels that contain a great deal of potential energy. Steam escapes and boiler explosions (typically BLEVEs) can and have in the past caused great loss of life. While variations in standards may exist in different countries, stringent legal, testing, training, care with manufacture, operation and certification is applied to ensure safety.

Failure modes may include:

- Over-pressurisation of the boiler.

- Insufficient water in the boiler causing overheating and vessel failure.

- Buildup of sediment and scale which cause local hot spots, especially in riverboats using dirty feed water.

- Pressure vessel failure of the boiler due to inadequate construction or maintenance.

- Escape of steam from pipework/boiler causing scalding.

Steam engines frequently possess two independent mechanisms for ensuring that the pressure in the boiler does not go too high; one may be adjusted by the user, the second is typically designed as an ultimate fail-safe. Such safety valves traditionally used a simple lever to restrain a plug valve in the top of a boiler. One end of the lever carried a weight or spring that restrained the valve against steam pressure. Early valves could be adjusted by engine drivers, leading to many accidents when a driver fastened the valve down to allow greater steam pressure and more power from the engine. The more recent type of safety valve uses an adjustable spring-loaded valve, which is locked such that operators may not tamper with its adjustment unless a seal is illegally broken. This arrangement is considerably safer.

Lead fusible plugs may be present in the crown of the boiler's firebox. If the water level drops, such that the temperature of the firebox crown increases significantly, the lead melts and the steam escapes, warning the operators, who may then manually suppress the fire. Except in the smallest of boilers the steam escape has little effect on dampening the fire. The plugs are also too small in area to lower steam pressure significantly, depressurizing the boiler. If they were any larger, the volume of escaping steam would itself endanger the crew.

Steam Cycle

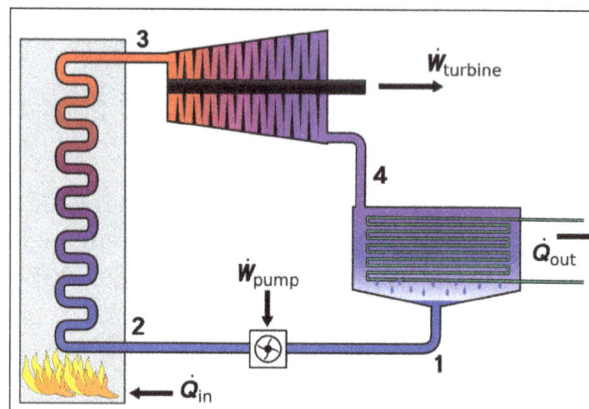

Flow diagram of the four main devices used in the Rankine cycle. 1). Feedwater pump 2). Boiler or steam generator 3). Turbine or engine 4). Condenser; where Q=heat and W=work. Most of the heat is rejected as waste.

The Rankine cycle is the fundamental thermodynamic underpinning of the steam engine. The cycle is an arrangement of components as is typically used for simple power production, and utilizes the phase change of water (boiling water producing steam, condensing exhaust steam, producing liquid water)) to provide a practical heat/power conversion system. The heat is supplied externally to a closed loop with some of the heat added being converted to work and the waste heat being removed in a condenser. The Rankine cycle is used in virtually all steam power production applications. In the 1990s, Rankine steam cycles generated about 90% of all electric power used throughout the world, including virtually all solar, biomass, coal and nuclear power plants. It is named after William John Macquorn Rankine, a Scottish polymath.

The Rankine cycle is sometimes referred to as a practical Carnot cycle because, when an efficient turbine is used, the TS diagram begins to resemble the Carnot cycle. The main difference is that heat addition (in the boiler) and rejection (in the condenser) are isobaric (constant pressure) processes in the Rankine cycle and isothermal (constant temperature) processes in the theoretical Carnot cycle. In this cycle, a pump is used to pressurize the working fluid which is received from the condenser as a liquid not as a gas. Pumping the working fluid in liquid form during the cycle requires a small fraction of the energy to transport it compared to the energy needed to compress the working fluid in gaseous form in a compressor (as in the Carnot cycle). The cycle of a reciprocating steam engine differs from that of turbines because of condensation and re-evaporation occurring in the cylinder or in the steam inlet passages.

The working fluid in a Rankine cycle can operate as a closed loop system, where the working fluid is recycled continuously, or may be an "open loop" system, where the exhaust steam is directly released to the atmosphere, and a separate source of water feeding the boiler is supplied. Normally water is the fluid of choice due to its favourable properties, such as non-toxic and unreactive chemistry, abundance, low cost, and its thermodynamic properties. Mercury is the working fluid in the mercury vapor turbine. Low boiling hydrocarbons can be used in a binary cycle.

The steam engine contributed much to the development of thermodynamic theory; however, the only applications of scientific theory that influenced the steam engine were the original concepts of harnessing the power of steam and atmospheric pressure and knowledge of properties of heat and steam. The experimental measurements made by Watt on a model steam engine led to the development of the separate condenser. Watt independently discovered latent heat, which was confirmed by the original discoverer Joseph Black, who also advised Watt on experimental procedures. Watt was also aware of the change in the boiling point of water with pressure. Otherwise, the improvements to the engine itself were more mechanical in nature. The thermodynamic concepts of the Rankine cycle did give engineers the understanding needed to calculate efficiency which aided the development of modern high-pressure and -temperature boilers and the steam turbine.

Efficiency

The efficiency of an engine cycle can be calculated by dividing the energy output of mechanical work that the engine produces by the energy put into the engine by the burning fuel.

The historical measure of a steam engine's energy efficiency was its "duty". The concept of duty was first introduced by Watt in order to illustrate how much more efficient his engines were over the earlier Newcomen designs. Duty is the number of foot-pounds of work delivered by burning one bushel (94 pounds) of coal. The best examples of Newcomen designs had a duty of about 7 million, but most were closer to 5 million. Watt's original low-pressure designs were able to deliver duty as high as 25 million, but averaged about 17. This was a three-fold improvement over the average Newcomen design. Early Watt engines equipped with high-pressure steam improved this to 65 million.

No heat engine can be more efficient than the Carnot cycle, in which heat is moved from a high-temperature reservoir to one at a low temperature, and the efficiency depends on the temperature difference. For the greatest efficiency, steam engines should be operated at the highest steam temperature possible (superheated steam), and release the waste heat at the lowest temperature possible.

The efficiency of a Rankine cycle is usually limited by the working fluid. Without the pressure reaching supercritical levels for the working fluid, the temperature range over which the cycle can operate is small; in steam turbines, turbine entry temperatures are typically 565 °C (the creep limit of stainless steel) and condenser temperatures are around 30 °C. This gives a theoretical Carnot efficiency of about 63% compared with an actual efficiency of 42% for a modern coal-fired power station. This low turbine entry temperature (compared with a gas turbine) is why the Rankine cycle is often used as a bottoming cycle in combined-cycle gas turbine power stations.

One principal advantage the Rankine cycle holds over others is that during the compression stage relatively little work is required to drive the pump, the working fluid being in its liquid phase at this point. By condensing the fluid, the work required by the pump consumes only 1% to 3% of the turbine (or reciprocating engine) power and contributes to a much higher efficiency for a real cycle. The benefit of this is lost somewhat due to the lower heat addition temperature. Gas turbines, for instance, have turbine entry temperatures approaching 1500 °C. Nonetheless, the efficiencies of actual large steam cycles and large modern simple cycle gas turbines are fairly well matched.

A steam locomotive – a GNR N2 Class No.1744 at Weybourne nr. Sheringham, Norfolk.

In practice, a reciprocating steam engine cycle exhausting the steam to atmosphere will typically have an efficiency (including the boiler) in the range of 1–10%, but with the addition of a condenser, Corliss valves, multiple expansion, and high steam pressure/temperature, it may be greatly improved, historically into the range of 10–20%, and very rarely slightly higher.

A steam-powered bicycle by John van de Riet, in Dortmund.

A modern, large electrical power station (producing several hundred megawatts of electrical output) with steam reheat, economizer etc. will achieve efficiency in the mid 40% range, with the most efficient units approaching 50% thermal efficiency.

It is also possible to capture the waste heat using cogeneration in which the waste heat is used for heating a lower boiling point working fluid or as a heat source for district heating via saturated low-pressure steam.

British horse-drawn fire engine with steam-powered water pump.

Electric Motor

An electric motor is a device used to convert electrical energy to mechanical energy. Electric motors are extremely important in modern-day life. They are used in vacuum cleaners, dishwashers, computer printers, fax machines, video cassette recorders, machine tools, printing presses, automobiles, subway systems, sewage treatment plants, and water pumping stations, to mention only a few applications.

Principle of Operation

The basic principle on which motors operate is Ampere's law. This law states that a wire carrying an electric current produces a magnetic field around itself.

In an AC motor, then, the current flows first in one direction through the wire loop and then reverses itself about 1/60 second later. This change of direction means that the magnetic field produced around the loop also changes once every 1/60 second. At one instant, one part of the loop is attracted by the north pole of the magnet, and at the next instant, it is attracted by the south pole of the magnet.

But this shifting of the magnetic field is necessary to keep the motor operating. When the current is flowing in one direction, the right hand side of the coil might become the south pole of the loop magnet. It would be repelled by the south pole of the outside magnet and attracted by the north pole of the outside magnet. The wire loop would be twisted around until the right side of the loop had completed half a revolution and was next to the north pole of the outside magnet.

If nothing further happened, the loop would come to a stop, since two opposite magnetic poles—one from the outside magnet and one from the wire loop—would be adjacent to (located next to) each other. And unlike magnetic poles attract each other. But something further does happen. The current changes direction, and so does the magnetic field around the wire loop. The side of the loop that was previously attracted to the north pole is now attracted to the south pole, and vice versa. Therefore, the loop receives another "kick," twisting it around on its axis in response to the new forces of magnetic attraction and repulsion.

Thus, as long as the current continues to change direction, the wire loop is forced to spin around on its axis. This spinning motion can be used to operate any one of the electrical appliances mentioned above.

DC motors. When electric motors were first invented, AC current had not yet been discovered. So the earliest motors all operated on DC current, such as the current provided by a battery.

Capacitor

A capacitor is a device for storing electrical energy. Capacitors are used in a wide variety of applications today. Engineers use large banks of capacitors, for example, to test the performance of an electrical circuit when struck by a bolt of lighting. The energy released by these large capacitors is similar to the lightning bolt. On another scale, a camera flash works by storing energy in a capacitor and then releasing it to cause a quick bright flash of light. On the smallest scale, capacitors are used in computer systems. A charged capacitor represents the number 1 and an uncharged capacitor a 0 in the binary number system used by computers.

How a capacitor stores energy A capacitor consists of two electrical conductors that are not in contact. The conductors are usually separated by a layer of insulating material known as a dielectric. Since air is a dielectric, an additional insulating material may not have to be added to the capacitor.

Think of a capacitor as consisting of two copper plates separated by 1 centimeter of air. Then imagine that electrical charge (that is, electrons) are pumped into one of the plates. That plate becomes negatively charged because of the excess number of electrons it contains. The negative charge on the first copper plate then induces (creates) a positive charge on the second plate.

As electrons are added to the first plate, one might expect a current to flow from that plate to the second plate. But the presence of the dielectric prevents any flow of electrical current. Instead, as more electrons are added to the first plate, it accumulates more and more energy. Adding electrons increases energy because each electron added to the plate has to overcome repulsion from other electrons already there. The tenth electron added has to bring with it more energy to add to the plate than did the fifth electron. And the one-hundredth electron will have to bring with it even more energy. As a result, as long as current flows into the first plate, it stores up more and more electrical energy.

Capacitors release the energy stored within them when the two plates are connected with each other. For example, just closing an electric switch between the two plates releases the energy stored in the first plate. That energy rushes through the circuit, providing a burst of energy.

The primary difference between a DC motor and an AC motor is finding a way to change the direction of current flow. In direct current, electric current always moves in the same direction. That means that the wire loop in the motor will stop turning after the first half revolution. Because the current is always flowing in the same direction, the resulting magnetic field always points in the same direction.

To solve this problem, the wire coming from the DC power source is attached to a metal ring cut in half, as shown in the figure. The ring is called a split-ring commutator. At the first moment the motor is turned on, current flows out of the battery, through the wire, and into one side of the commutator. The current then flows into the wire loop, producing a magnetic field.

Once the loop begins to rotate, however, it carries the commutator with it. After a half turn, the ring reaches the empty space in the two halves and then moves on to the second half of the commutator. At that point, then, current begins to flow into the opposite side of the loop, producing the same effect achieved with AC current. Current flows backward through the loop, the magnetic field is reversed, and the loop continues to rotate.

References

- Erjavec, Jack (2010). Automotive Technology: A Systems Approach. Clifton Park, NY USA: Delmar, Cengage Learning. Pp. 226–227. ISBN 978-1428311497. LCCN 2008934340. Retrieved 2014-02-09

- Internal-combustion-engine, entry: newworldencyclopedia.org, Retrieved 5 March, 2019

- Hunt, Phil; mckay, Malcolm; Wilson, Hugo; Robinson, James (2012), Duckworth, Mick (ed.), Motorcycle: The Definitive Visual History, DK Publishing, Penguin Group, pp. 126, 210, ISBN 9781465400888

- Gasoline-engine, technology: britannica.com, Retrieved 6 April, 2019

- Rik D Meininger et al.: Knock criteria for aviation diesel engines, International Journal of Engine Research, Vol 18, Issue 7, 2017, doi/10.1177

- Rotary-engine, encyclopedia: energyeducation.ca, Retrieved 7 May, 2019

- Andrew Roberts (July 10, 2007). "Peugeot 403". The 403, launched half a century ago, established Peugeot as a global brand. The Independent, London. Retrieved February 28, 2019

3

Engine Management Systems and Control Module

The electronic control unit which controls a various actuators on an internal combustion engine to derive optimal engine performance is known as an engine management system. There are various engine management systems such as Trionic and Trionic T5.5 which have been described and thoroughly explained in this chapter.

Engine Management System

An Engine Management System is now very common or even required on both high end and regular vehicles. Its also common for people to use the term Engine Control Unit (ECU) in place of engine management system. The engine management system is basically an electronic control unit (ECU) which receives signals from various sensors, make calculations and sends output signals to carry out various functions and operations within and around the engine. The main reason for a proper engine management system is to reduce emissions and achieve better fuel economy. Performance is also increased but vehicles before management systems used to perform well for their time but an engine management system achieves the performance figures with better fuel economy and less emissions.

The benefits of an engine management system will be the same as the benefits of a fuel injected engine as fuel injection is controlled by an engine management system but there are some other benefits when a sophisticated and powerful system is used. Some of the other benefits of an engine control unit is that it can alter certain operations in the same engine to produce a different result. For example an automotive marque can have the same 5.0 liter engine in two seperate models but the ECU on both engines can be tuned differently to allow one of the engines to have a smooth luxury feel and the other to have a sporty character. More advanced uses of an engine control unit can facilitate both the comfort and sporty characters of the engine in a single vehicle and usually combines with other features which allows the vehicle to have driving modes. Some of the common sensors which sends information to the engine management system are the crankshaft position sensor, camshaft position sensor, mass airflow sensor, air temperature, coolant temperature, throttle position, knock sensor, oxygen sensors and many others. Some of the common places where the engine management system sends instructions and information to will be the electronic injectors, spark plugs, ignition coils, cooling fans, engine immobiliser and displays such as but not limited to the tachometer and the temperature guage. All of these sensors that send information and all of the areas where the ECU sends commands to adds up to alot. The processor of an engine management system usually has to perform thousands of operations every second but the management systems are built to handle more operations than the engine will give it.

Trionic

Trionic is an innovative engine management system developed by Saab Automobile, consisting of an engine control unit (ECU) that controls 3 engine aspects:

- Ignition timing,

- Fuel injection,

- Acts as a boost controller.

Hence the numerical prefix 'tri-' in Trionic. 'Ion' comes from the fact that it uses ion current measured by the spark plugs between combustion events as a sensor for knock, misfire and synchronization detection. With the ion sensing system, the ion current stream developed due to combustion can be deduced by monitoring the secondary current of the ignition coil. Using the value and wave shape of the current, after the actual spark event, the quality of the actual combustion process is determined, thus allowing the engine control unit to optimize the timing of the spark for the best engine performance while keeping emissions low on a much wider range of rpms.

Since Trionic 7, the throttle and thereby the air charge is electronically controlled, but the name "Trionic" was not changed accordingly.

Trionic T5.5

Trionic T5.5 is an engine management system in the Saab Trionic range. It controls ignition, fuel injection and turbo boost pressure. The system was introduced in the 1994 Saab 900 with B204L engine.

Changes

Since 1994 a number of changes have occurred.

- 1995: Four wire oxygen sensor, electronic heat plates in intake manifold (not in US and CA markets). K line is connected via VSS (Vehicle Security System) to enable immobilizing (certain markets). Vacuum pump for the vacuum servo assisted brake system with some control from Trionic is used on automobiles with automatic transmission.

- 1996: OBD II diagnostics on US and CA markets, which means two lambda probes.

- 1996: Leakage diagnostics of the EVAP system on the OBD II variant.

- 1997: Heat plates are removed.

- 1998: (Saab 9-3). K-line is connected via MIU (Main Instrument Unit) to enable immobilizing from TWICE (Theft Warning Integrated Central Electronics) (not in software for markets: US and CA). Fuel pump relay is electrically supplied from main relay. Request signal for Air Condition is feed from MIU. Electrical pre heating on oxygen sensor is supplied from main relay. Requested boost pressure is raised somewhat on automobiles with manual gearbox. SID message when leakage in EVAP-system is confirmed, applicable in On-Board Diagnostics II variants.

- 1998: Two new engine variants; B204R and B204E, B204E were available with manual gearbox only and demanded high octane gasoline to deliver the stated torque. B204E is lacking boost pressure control, this engine wasn't available on US and CA markets. On the Swedish market automobiles is equipped with the B204E engine, OBD II diagnostics and ORVR (On board Refuelling Vapour Recovery system), a system that makes sure that the gasoline vapour doesn't escape into the surrounding air during refuelling.

Description

Saab Trionic's ignition system consists of an ignition cassette with four ignition coils, one for each spark plug. The ignition system is capacitive. The spark plugs are used as sensors to detect combustion and pre-ignition/pinging. This renders camshaft position detector and knock sensor redundant. This function also enables effective detection of misfires, which is an OBD II demand. The fuel injection is fully sequential and is dependent on the MAP (Manifold Absolute Pressure). Boost pressure control (L and R engines) utilises a solenoid valve pneumatically connected to the turbocharger's waste gate.

The system was fitted on models Saab 900, Saab 9000 and Saab 9-3. This information is however most accurate for the SAAB 900.

Fuel

Fuel Injector Valves

The fuel injector valves are of a solenoid type with needle and seat. They are opened by a current flowing through the injector's coil and are closed by a strong spring when the current is switched off. To ensure as optimal combustion as possible and with that lower exhaust emission the injectors are equipped with four holes, which gives a good distribution of the fuel. The squirts of fuel are very exact positioned (two jets on the backside on each inlet valve). This put very high demands on the fixation of the injectors. To secure this fixation the injectors are fixed in pairs by a special retainer between cylinders 1 – 2 and 3 – 4. The injectors are electrically supplied from the main relay, while the ECU grounds the injectors.

Fuel Injection

Pre-injection

When the ignition is switched on, the main relay and fuel pump relay are activated during a few seconds. As soon as the ECU gets the cranking signal (from the crankshaft sensor) it initiates a coolant temperature dependent fuel injection with all four injectors simultaneously which ensures a fast engine start. If the engine is started and shortly after is switched off a new pre-injection is initiated after the ignition has been switched off for 45 seconds.

Calculating of Injection Time

To decide how much fuel needs to be injected into each intake runner the ECU calculates the air mass that had been drawn into the cylinder. The calculation makes use of the cylinder volume (the B204 engine has a displacement of 0.5 litres per cylinder). That cylinder volume holds equal amount of air which has a density and thus a certain mass. The air density is calculated using the absolute pressure and temperature in the intake manifold. The air mass for combustion has now been calculated and that value is divided by 14.7 (stoichiometric relation for gasoline mass to air mass) to determine the required fuel mass for each combustion to inject. Since the flow capacity of the injector and the density of the fuel (pre programmed values) are known, the ECU can calculate the duration of the injection.

Using the oxygen sensor 1 the injection duration is corrected so stoichiometric combustion is obtained. When hard acceleration occurs, the lambda correction is masked and Wide Open Throttle (WOT) enrichment occurs for maximum performance. When opening the throttle, acceleration enrichment occurs and when closing the throttle deceleration emaciation occurs. During a cold start and warm up, before lambda correction is activated, coolant temperature dependable fuel enrichment occurs. With a warm engine and normal battery voltage the duration of injection varies between 2,5 ms at idle and approx. 18 – 20 ms at full torque.

Lambda Correction

The catalyst requires that the fuel/air mixture is stoichiometric. This means that the mixture is neither rich or lean, it is exactly 14,7 kg air to 1 kg gasoline (Lambda=1). That is why the system is equipped with an oxygen sensor in the forward part of the exhaust system. The sensor is connected to pin 23 in the ECU and is grounded in the ECU via pin 47. The exhaust fumes pass the oxygen sensor. The content of oxygen in the exhaust fumes is measured through a chemical reaction, this results in an output voltage. If the engine runs rich (Lambda lower than 1) the output voltage would be more than 0.45 V and if the engine runs lean (Lambda higher than 1) the output voltage would be less than 0.45 V. The output voltage swings about 0.45 V when Lambda passes 1. The ECU continuously corrects the injection duration so that Lambda=1 is always met. To be able to function the oxygen sensor needs to be hot, this requirement is meet by electrically pre heat the sensor. The pre heating element is fed by B+ via fuse 38 and the main relay, the sensor is grounded in the ECU via pin 50. The ECU estimates the temperature on the exhaust gases (EGT) on the basis of the engine load and the engines RPM. At high EGT the electrical pre heating is disconnected. The lambda correction is masked during the engines first 640 revolutions after start if the coolant temperature exceeds 18 °C (64F) at load ranges over idle and under WOT or 32 °C (90F) at idle.

Adaptation

The ECU calculates the injection duration on basis of MAP and intake temperature. Injection duration are then corrected by multiplication of a correction factor, which is fetched from main fuel matrix and is dependable on MAP and RPM. The need to correct the injection duration is due that the volumetric efficiency of the cylinder is dependent on the engines RPM. The last correction is made with the lambda correction, this results in a stoichiometric combustion (Lambda=1). The lambda correction is allowed to adjust the calculated injection duration by ±25%. The ECU can change the correction factors in the main fuel matrix on basis of the lambda correction, this ensures good driveability, fuel consumption and emissions when lambda correction isn't activated. This is called Adaptation.

Pointed Adaptation

If the ECU calculates the injection duration to 8 ms but the lambda correction adjusts it to 9 ms due low fuel pressure the ECU will "learn" the new injection duration. This is done by changing the correction factor for that particular RPM and load point in the main fuel matrix to a new correction factor resulting in 9 ms injection duration. The correction factor in this example will be raised by 9/8 (+12%). The pointed adaptation can change the points in the main fuel matrix by ±25%. Adaptation occurs every fifth minute and takes 30 seconds to finish, the criteria for the adaptation are: Lambda correction is activated and the coolant temperature is above 64 °C (147F). During the adaptation the ventilation valve on the carbon canister is held close.

Global Adaptation

The global adaptation on OBDII variants occurs during driving; on non OBDII variants the global adaptation occurs 15 minutes after engine shut down. When the engine is inside a defined load and RPM range (60 – 120 kPa and 2000 – 3000 RPM) no pointed adaptation will occur all points in the fuel matrix will be changed instead by a multiplication factor. Global adaptation can change the points in the main fuel matrix by ±25% (Tech2 shows ±100%). Adaptation occurs every fifth minute and takes 30 seconds to finish, the criteria for the adaptation are: Lambda correction is activated and the coolant temperature is above 64 °C (147F). During the adaptation the ventilation valve on the carbon canister is held close.

Fuel Cut

With fully closed throttle and engine RPM over 1900 RPM and with third, fourth and fifth gear a fuel cut will occur after a small delay (some second). On automobiles with automatic transmission fuel cut is active in all stages. The injectors are reactivated when the RPM hits 1400 RPM.

Fuel Consumption Measurement

The wire from the ECU to the third injector is also connected to the main instrument. The main instrument calculates the fuel consumption based on the injection pulses duration. The fuel consumption is used to help getting an accurate presentation of the fuel level in the fuel tank and to calculate average fuel consumption in SID.

Turbo Boost Pressure

Basic Charging Pressure

Basic charging pressure is fundamental for Automatic Performance Control (APC). Basic charging pressure is mechanically adjusted on the actuators pushrod between the actuator and the waste gate. At to low basic charging pressure the engine doesn't revs up as expected when the throttle is opened quickly. At to high basic charging pressure a negative adaptation occurs and maximum charging pressure cannot be achieved. In addition there is a substantial risk of engine damage since the charging pressure can't be lowered enough when regulating with attention to pre ignition/pinging. Basic charging pressure shall be 0,40 ±0,03 bar (5,80 ±0,43 PSI). After adjustment the push rod must have at least two turns (2 mm) pre tension when connecting to the waste gate lever. The purpose with that is to make sure that the waste gate is held close when not affected. On new turbo chargers the basic charging pressure tends to be near or spot on the upper tolerance when the pre tension is two turns. The pre tension may never be lesser than two turns (2 mm). When checking the basic charging pressure it shall be noted that the pressure decreases at high RPM and increases at low outside temperatures.

Charging Pressure Regulation

Charging pressure regulation utilises a two coiled three way solenoid valve pneumatically connected with hoses to the turbo charger's waste gate, the turbo chargers outlet and the compressor's inlet. The solenoid valve is electrically supplied from +54 via fuse 13 and is controlled by the ECU via its pin 26 and pin 2. The control voltage is pulse width modulated (PWM) at 90 Hz below 2500 RPM and 70 Hz above 2500 RPM. The rationale for this change is to avoid resonance phenomena in the pneumatic hoses. By grounding pin2 longer than pin 26 the charging pressure is decreased and vice verse, when pin 26 is grounded longer than pin 2 the charging pressure is increased. To be able to regulate the charging pressure the ECU must at first calculate a requested pressure, a pressure value that the system must strive for. This is done by taking a pre programmed value (matrix of values established in respect of RPM and throttle opening). At WOT the pressure values for each RPM are selected to make sure that the engine gets the requested torque.

When one or both of the following criteria are met, a limitation of the charging pressure is set.

- In first, second and reverse gear there is an RPM dependable maximum value. The ECU calculates which gear that is in use by comparing the speed of the automobile and the engines RPM.

- When pre ignition/pinging occurs a maximum charge pressure is set on the basis of a mean value from each cylinders retarding of the ignition.

- One or both of the following criteria initiates a lowering of the charging boost pressure to basic boost pressure.

- When the brake pedal is pressed down and pin 15 on the ECU is supplied with battery voltage.

- Certain fault codes is set (Faulty throttle position sensor (TPS), pressure sensor, pre ignition/pinging signal or charging pressure regulation) or low battery voltage.

Computing and Adaptation

When the required charge pressure has finally been calculated it is converted to the PWM signal that controls the solenoid valve, The ECU then controls that the actual pressure (measured by the pressure sensor) corresponds with the required pressure. If needed the PWM is fine tuned by multiplication of a correction factor. The correction factor (adaptation) is then stored in the memory of the ECU and is always used in the calculation of the PWM signal. The rationale with this is to make sure that the actual pressure as soon as possible will be equal to the required after a change of the load has occurred.

Ignition Timing

Ignition Cassette

The red ignition cassette used with Trionic 5 is mounted on the valve cover on top of the spark plugs. The ignition cassette houses four ignition coils/transformers whose secondary coil is direct connected to the spark plugs. The cassette is electrically supplied with battery voltage from the main relay (B+) and is grounded in an earth point. When the main relay is activated the battery voltage is reformed to 400 V DC which is stored in a capacitor. 400 V voltage is connected to one of the poles of the primary coil in the four spark coils. To the ignition cassette there are four triggering lines connected from the Trionic ECU, pin 9 (cyl. 1), pin 10 (cyl. 2), pin 11 (cyl. 3) and pin 12 (cyl. 4). When the ECU is grounding pin 9, the primary coil for the first cylinder is grounded (via the ignition cassettes B+ intake) and 400 V is transformed up to a maximum of 40 kV in the secondary coil for cyl. 1. The same procedure is used for controlling the ignition timing of the rest of the cylinders.

Ignition Regulation

At start the ignition point is 10° BTDC. To facilitate start when coolant temperature is below 0°C the ECU will ground each trigger line 210 times/second between 10° BTDC and 20° ATDC, at which a "multi spark" will appear. The function is active up to an engine speed of 900 RPM. At idle a special ignition matrix is utilised. Normal ignition point is 6°-8° BTDC. If the engine stalls e.g. cooling fan activation the ignition point is advanced up to 20 ° BTDC in order to increase the engines torque to restore the idle RPM. In the same way the ignition is retarded if the engines RPM is increased. When the TPS senses an increase in throttle opening the ECU leaves the idle ignition timing map and regulates the ignition timing in respect of load and engine speed.

During engine operations the Ignition cassette continuously monitors the ion currents in the cylinders and sends a signal to the Trionic ECU, pin 44, in an event of knocking. The logic for this function rests solely in the ignition cassette and is adaptive to be able to handle disturbing fuel additives. The Trionic ECU is well aware which cylinder that has ignited and could hence cope with the information feed through one pin. The signal to pin 44 and ion current in the combustion chamber is related to each other, when this signal reaches a certain level the ECU interprets this as a knocking event and firstly lowering the ignition advance by 1,5° on this cylinder. If the knocking is repeated the ignition advance is lowered further 1,5 ° up to 12°. In case of the same lowering of the ignition timing advance in all cylinders the ECU adds a small amount of fuel to all cylinders. If knocking occurs when the MAP is over 140 kPa the knocking is regulated by switching both fuel injection matrix and ignition advance matrix. If this is not sufficient the charging pressure is lowered. This purpose of this procedure is to maintain good performance. If the signal between the ignition cassette and the ECU is lost, the charging pressure is lowered to basic charging pressure and the ignition timing advance is lowered 12° when it exist a risk of knocking due to engine load.

Combustion Signals

The Trionic system lacks a camshaft position sensor. This sensor is normally a prerequisite for a sequential pre ignition/pinging regulation and fuel injection. Saab Trionic must decide whether cylinder one or cylinder four ignites when the crank shaft position sensor indicates that cylinder one and four is at TDC. This is done by help of ionization current. One of the poles of the secondary coil of the spark coils is connected to the spark plugs in an ordinary manner. The other pole isn't grounded directly but connected to an 80 V voltage. This means that an 80 V voltage is applied across the spark gap of the spark plugs, except when the spark is fired. When combustion has occurred the temperature in the combustion chamber is very high. The gases are formed as ions and start to conduct electric current. This results in a current flowing in the spark plug gap (without resulting in a spark). The ionisation current is measured in pairs, cylinder one and two is one pair and cylinder three and four in the other pair. If combustion occurs in cylinder one or two the ignition cassette sends battery voltage (B+) pulse to the ECU, pin 17. If the combustion takes place in cylinder three or four the B+ pulse is fed to pin 18 in ECU. If the crankshaft position sensor indicates that cylinders one and four are at TDC and a B+ pulse enters the ECU via pin 17 simultaneously, then the ECU know that it is cylinder one that has ignited. Upon starting, the ECU doesn't know which cylinder is in compression phase, hence ignition is initiated in both cylinders one and four and 180 crank shaft degrees later sparks in cylinder two and three are fired. As soon as the combustion signals enters the ECU via pin 17 and pin 18 the ignition and fuel injection is synchronised to the engines firing order. The combustion signals are also used to detect misfires.

Heat Plates

Heat plates are used to lower the warm up emissions. They vaporize the injected fuel before it is drawn/forced into the cylinders and consequently reduce the need for added fuel in the A/F mixture in the warm up phase thus reducing the emissions. At engine start and coolant temperature lower than +85°C Pin 29 on ECU is grounded and a relay in the engine compartment are activated and closes the electrical circuit for the Heat Plates. The circuit is protected by a 40 A MAXI fuse. When the coolant temperature is warmer than +85°C or four minutes has passed the Heat Plates are switched off.

To compensate for the increased air resistance in the intake, engines fitted with Heat Plates have a slightly adjusted charge pressure, Approximately: +0.2 bar, this means that LPT models with heat plates have a solenoid valve to raise the charging pressure above basic charging pressure.

In case of a Heat Plate-failure the car may have drivability problems due condensed fuel in the intake during cold engine operations. This condensed fuel is compensated in engines without Heat Plates by enriching the A/F mixture.

The heat plates are activated by software, which enables different algorithms to use the plates and to compensate for the intake restriction caused by the plates' presence.

Other Features

Shift Up Lamp

The Shift Up lamp can be found on OBD II cars. The lamp helps the driver to drive economically.

The lamp is supplied by ignition power (+15) and is grounded in the Trionic ECU, pin 55. The Shift Up Lamp is lit when the ignition is turned on for three seconds to test the circuit. During normal driving the lamp is lit when reaching a specific RPM while driving at light loads. At wide-open-throttle the Shift Up lamp is lit when the RPM is near 6000 RPM. The lamp does not light in fifth gear. The light is lit at a higher RPM when the engine is cold to promote a quicker warm up.

AMC Computerized Engine Control

The Computerized Engine Control or Computerized Emission Control (CEC) system is an engine management system designed and used by American Motors Corporation (AMC) and Jeep on 4- and 6-cylinder engines of its own manufacture from 1980 to 1990.

Starting with the 1986 model year, the AMC straight-4 engines used a throttle body injection (TBI) or single-point, fuel injection system with a new fully computerized engine control. In addition to cycling the fuel injector (pulse-width time, on–off), the engine control computer also determined the ignition timing, idle speed, exhaust gas recirculation, etc.

Operation

CEC was unique in that almost all of its sensors and actuators were digital; instead of the usual analog throttle position, coolant temperature, intake temperature and manifold pressure sensors, it used a set of fixed pressure- and temperature-controlled switches (as well as a wide-open throttle switch on the carburetor) to control fuel mixture and ignition timing. The only analog sensor in the system was the oxygen sensor. In other respects, it was a typical "feedback" carburetor system of the early-1980s, using a stepper motor to control fuel mixture and a two-stage "Sole-Vac" (which used a solenoid for one stage, and a vacuum motor for the other) to control idle speed. The CEC also controlled ignition timing using information from the fuel-control section and an engine knock sensor on the intake manifold.

The CEC module itself (the most common version of which is the "AMC MCU Super-D") was manufactured for AMC by Ford Motor Company, and worked with a Duraspark ignition system. Although built by Ford, the CEC module is not related to the Ford EEC systems internally.

Starting with the 1983 model year, the 258 cu in (4.2 L) I6 engine featured the MCU-Super D electronics and a "pulse-air" injection system for emissions control, as well as an increased compression ratio, from 8.6:1 to 9.2:1.

Maintenance

The system uses a maze of emissions vacuum hoses. Because of the many vacuum-driven components and electrical connections in the system, CEC-equipped engines have developed a reputation of being hard to tune. American Motors issued a Technical Service Bulletin to diagnose low or inconsistent engine idle speeds on 1980 through 1988 AMC Eagle automobile.

The 49-state model of the CEC has no on-board diagnostic system, making it difficult to monitor the computer's operation without a breakout box, and the Carter BBD carburetor on most CEC-equipped models has problems with its idle circuit clogging, causing a rough idle and stalling. In places where emissions testing is not required, a popular modification is to bypass the computer and disable the BBD's Idle Servo, or replace the BBD with a manually tuned carburetor. Several vendors (including Chrysler and Edelbrock) offer retrofit kits that replace the CEC and the carburetor with fuel injection.

Engine Control Module

An ECU from a 1996 Chevrolet Beretta.

An engine control unit (ECU), also commonly called an engine control module (ECM), is a type of electronic control unit that controls a series of actuators on an internal combustion engine to ensure optimal engine performance. It does this by reading values from a multitude of sensors within the engine bay, interpreting the data using multidimensional performance maps (called lookup tables), and adjusting the engine actuators. Before ECUs, air-fuel mixture, ignition timing, and idle speed were mechanically set and dynamically controlled by mechanical and pneumatic means.

If the ECU has control over the fuel lines, then it is referred to as an electronic engine management system (EEMS). The fuel injection system has the major role to control the engine's fuel supply. The whole mechanism of the EEMS is controlled by a stack of sensors and actuators.

Workings

Control of Air–fuel Ratio

Most modern engines use some type of fuel injection to deliver fuel to the cylinders. The ECU determines the amount of fuel to inject based on a number of sensor readings. Oxygen sensors tell the ECU whether the engine is running rich (too much fuel or too little oxygen) or running lean (too much oxygen or too little fuel) as compared to ideal conditions (known as stoichiometric). The throttle position sensor tells the ECU how far the throttle plate is opened when the accelerator (gas pedal) is pressed down. The mass air flow sensor measures the amount of air flowing into the engine through the throttle plate. The engine coolant temperature sensor measures whether the engine is warmed up or cool. If the engine is still cool, additional fuel will be injected.

Air–fuel mixture control of carburetors with computers is designed with a similar principle, but a mixture control solenoid or stepper motor is incorporated in the float bowl of the carburetor.

Control of Idle Speed

Most engine systems have idle speed control built into the ECU. The engine RPM is monitored by the crankshaft position sensor which plays a primary role in the engine timing functions for fuel injection, spark events, and valve timing. Idle speed is controlled by a programmable throttle stop or an idle air bypass control stepper motor. Early carburetor-based systems used a programmable throttle stop using a bidirectional DC motor. Early throttle body injection (TBI) systems used an idle air control stepper motor. Effective idle speed control must anticipate the engine load at idle.

A full authority throttle control system may be used to control idle speed, provide cruise control functions and top speed limitation. It also monitors the ECU section for reliability.

Control of Variable Valve Timing

Some engines have variable valve timing. In such an engine, the ECU controls the time in the engine cycle at which the valves open. The valves are usually opened sooner at higher speed than at lower speed. This can increase the flow of air into the cylinder, increasing power and fuel economy.

Electronic Valve Control

Experimental engines have been made and tested that have no camshaft, but have full electronic control of the intake and exhaust valve opening, valve closing and area of the valve opening. Such engines can be started and run without a starter motor for certain multi-cylinder engines equipped with precision timed electronic ignition and fuel injection. Such a *static-start* engine would provide the efficiency and pollution-reduction improvements of a mild hybrid-electric drive, but without the expense and complexity of an oversized starter motor.

The first production engine of this type was invented in 2002 and introduced in 2009 by Italian automaker Fiat in the Alfa Romeo MiTo. Their Multiair engines use electronic valve control which dramatically improve torque and horsepower, while reducing fuel consumption as much as 15%. Basically, the valves are opened by hydraulic pumps, which are operated by the ECU. The valves can open several times per intake stroke, based on engine load. The ECU then decides how much fuel should be injected to optimize combustion.

At steady load conditions, the valve opens, fuel is injected, and the valve closes. Under a sudden increase in throttle, the valve opens in the same intake stroke and a greater amount of fuel is injected. This allows immediate acceleration. For the next stroke, the ECU calculates engine load at the new, higher RPM, and decides how to open the valve: early or late, wide-open or half-open. The optimal opening and timing are always reached and combustion is as precise as possible. This, of course, is impossible with a normal camshaft, which opens the valve for the whole intake period, and always to full lift.

The elimination of cams, lifters, rockers, and timing set reduces not only weight and bulk, but also friction. A significant portion of the power that an engine actually produces is used up just driving the valve train, compressing all those valve springs thousands of times a minute.

Once more fully developed, electronic valve operation will yield even more benefits. Cylinder deactivation, for instance, could be made much more fuel efficient if the intake valve could be opened on every downstroke and the exhaust valve opened on every upstroke of the deactivated cylinder or "dead hole". Another even more significant advancement will be the elimination of the conventional throttle. When a car is run at part throttle, this interruption in the airflow causes excess vacuum, which causes the engine to use up valuable energy acting as a vacuum pump. BMW attempted to get around this on their V-10 powered M5, which had individual throttle butterflies for each cylinder, placed just before the intake valves. With electronic valve operation, it will be possible to control engine speed by regulating valve lift. At part throttle, when less air and gas are needed, the valve lift would not be as great. Full throttle is achieved when the gas pedal is depressed, sending an electronic signal to the ECU, which in turn regulates the lift of each valve event, and opens it all the way up.

Programmability

A special category of ECUs are those which are programmable; these units can be reprogrammed by the user.

When modifying an engine to include aftermarket or upgrade components, stock ECUs may or may not be able to provide the correct type of control for the application(s) in which the engine may be used. To accommodate for engine modifications, a programmable ECU can be used in place of the factory-shipped ECU. Typical modifications that may require an ECU upgrade can include turbocharging, supercharging, or both, a naturally aspirated engine; fuel injection or spark plug upgrades, exhaust system modifications or upgrades, transmission upgrades, and so on. Programming an ECU typically requires interfacing the unit with a desktop or laptop computer; this interfacing is required so the programming computer can send complete engine tunings to the engine control unit as well as monitor the conditions of the engine in realtime. Connection typically used in this interface are either USB or serial.

By modifying these values while monitoring the exhausts using a wide band lambda probe, engine tuning specialists can determine the optimal fuel flow specific to the engine speed's and throttle position. This process is often carried out at a engine performance facility. A dynamometer is typically found at these locations; these devices can provide engine tuning specialist useful information such as engine speed, power output, torque output, gear change events, and so on. Tuning specialists often utilize a chassis dynamometer for street and other high performance applications.

Engine tuning parameters may include fuel injection volume, throttle-fuel volume mapping, gear shift mapping, and so forth. While the mentioned parameters are common, some ECUs may provide other variables in which a tuning software could potentially modify. These parameters include:

- Anti-lag,

- Closed loop Lambda: Lets the ECU monitor a permanently installed lambda probe and modify the fueling to achieve the targeted air/fuel ratio desired. This is often the stoichiometric (ideal) air fuel ratio, which on traditional petrol (gasoline) powered vehicles this air-to-fuel ratio is 14.7:1. This can also be a much richer ratio for when the engine is under high load, or possibly a leaner ratio for when the engine is operating under low load cruise conditions for maximum fuel efficiency,

- Gear control,

- Ignition timing,

- Launch control,

- Fuel pressure regulator,

- Rev limiter,

- Staged fuel injection,

- Transient fueling: Tells the ECU to add a specific amount of fuel when throttle is applied. This is referred to as "acceleration enrichment",

- Variable cam timing,

- Wastegate control,

- Water temperature correction: Allows for additional fuel to be added when the engine is cold, such as in a winter cold-start scenario or when the engine is dangerously hot, to allow for additional cylinder cooling (though not in a very efficient manner, as an emergency only).

A race-grade ECU is often equipped with a data logger to record all sensor data for later analysis. This can be useful for identifying engine stalls, misfires or other undesired behaviors during a race. The data logger usually has a capacity between 0.5 and 16 megabytes.

In order to communicate with the driver, a race ECU can often be connected to a "data stack", which is a simple dashboard presenting the driver with the current RPM, speed and other basic engine data. These data stacks, which are almost always digital, talk to the ECU using one of several protocols including RS-232 or CANbus. Information is then relayed through the Data Link interface that usually located on the underneath of the steering column.

Sensors and Actuators

Sensors for air flow, pressure, temperature, speed, exhaust oxygen, knock and crank angle position sensor makes a very vital impact in EEMS sensors. MAP: Manifold Absolute pressure; IAT: Intake Air Tempareture; MAF: Mass of Air Flow; CKP: Crank Shaft Position; CMP: CAM Shaft position; ECT: Engine coolant tempareture; O_2: Oxygen sensor; TP: throtle position; VSS: vehicle speed sensor; Knock sensor APP: Acceleration pedal position; Refrigrant sensor.

Modern Design

Modern ECUs use a microprocessor which can process the inputs from the engine sensors in real-time. An electronic control unit contains the hardware and software (firmware). The hardware consists of electronic components on a printed circuit board (PCB), ceramic substrate or a thin laminate substrate. The main component on this circuit board is a microcontroller chip (MCU). The software is stored in the microcontroller or other chips on the PCB., typically in EPROMs or flash memory so the CPU can be re-programmed by uploading updated code or replacing chips. This is also referred to as an (electronic) Engine Management System (EMS).

Sophisticated engine management systems receive inputs from other sources, and control other parts of the engine; for instance, some variable valve timing systems are electronically controlled, and turbocharger waste gates can also be managed. They also may communicate with transmission control units or directly interface electronically controlled automatic transmissions, traction control systems, and the like. The Controller Area Network or CAN bus automotive network is often used to achieve communication between these devices.

Modern ECUs sometimes include features such as cruise control, transmission control, anti-skid brake control, and anti-theft control, etc.

General Motors' (GM) first ECUs had a small application of hybrid digital ECUs as a pilot program in 1979, but by 1980, all active programs were using microprocessor based systems. Due to the large ramp up of volume of ECUs that were produced to meet the Clean Air Act requirements for 1981, only one ECU model could be built for the 1981 model year. The high volume ECU that was installed in GM vehicles from the first high volume year, 1981, onward was a modern microprocessor based system. GM moved rapidly to replace carburation with fuel injection as the preferred method of fuel delivery for vehicles it manufactured. This process first saw fruition in 1980 with fuel injected Cadillac engines, followed by the Pontiac 2.5L I4 "Iron Duke" and the Chevrolet 5.7L V8 L83 "Cross-Fire" engine powering the Chevrolet Corvette in 1982. The 1990 Cadillac Brougham powered by the Oldsmobile 5.0L V8 LV2 engine was the last carbureted passenger car manufactured for sale in the North American market (a 1992 Volkswagen Beetle model powered by a carbureted engine was available for purchase in Mexico but not offered for sale in the United States or Canada) and by 1991 GM was the last of the major US and Japanese automakers to abandon carburetion and manufacture all of its passenger cars exclusively with fuel injected engines. In 1988 Delco (GM's electronics division), had produced more than 28,000 ECUs per day, making it the world's largest producer of on-board digital control computers at the time.

Other Applications

Such systems are used for many internal combustion engines in other applications. In aeronautical applications, the systems are known as "FADECs" (Full Authority Digital Engine Controls). This kind of electronic control is less common in piston-engined light fixed-wing aircraft and helicopters than in automobiles. This is due to the common configuration of a carbureted engine with a magneto ignition system that does not require electrical power generated by an alternator to run, which is considered a safety advantage.

Symptoms of a Bad or Failing Engine Control Module

The ECM plays a crucial role in newer vehicles, where many of the essential functions of the car are controlled by the ECM. When the ECM has any issues, it can cause all sorts of problems with the vehicle, and in some cases even render it undrivable. A bad or failing ECM may produce any of the following 5 symptoms to alert the driver of a potential problem.

Check Engine Light Turns On

An illuminated Check Engine Light is one possible symptom of a problem with the ECM. The Check Engine Light usually illuminates when the computer detects a problem with any of its sensors or circuits. There are cases, however, where the ECM illuminates a Check Engine Light mistakenly, or when there is no issue present. Have a mechanic scan the computer for trouble codes to diagnose whether the issue is with the ECM, or elsewhere on the vehicle.

Engine Stalling or Misfiring

Erratic engine behaviour may also indicate a bad or failing ECM. A faulty computer may cause the vehicle to intermittently stall or misfire. The symptoms might come and go and not appear to have any sort of pattern to their frequency or severity.

Engine Performance Issues

Engine performance issues are another symptom of a possible problem with the ECM. If the ECM has any issues, it may throw off the timing and fuel settings of the engine, which can negatively affect performance. A faulty ECM may cause the vehicle to experience a reduction in fuel efficiency, power, acceleration.

Car not Starting

A bad ECM may lead to a vehicle that won't or is difficult to start. If the ECM fails completely, it will leave the vehicle without engine management control, and will not start or run as a result. The engine may still crank, but it will not be able to start without the vital inputs from the computer. Car-starting problems are not exclusively caused by the ECM, so it's best to get a complete diagnosis by a professional technician to accurately determine the cause.

Poor Fuel Economy

Bad fuel economy may occur from a failing ECM. A faulty ECM disallows your engine from knowing how much fuel to burn in the combustion process. Typically, the vehicle consumes more fuel than it should in this situation. You'll end up paying more for gas than you would with a functioning ECM.

The ECM plays a vital role in engine performance. Any issues with it can cause major problems with the overall functionality of the car. As the computer systems found on modern vehicles are quite sophisticated and complicated, they can also be difficult to diagnose. For this reason, if you suspect that your vehicle's ECM is having an issue, have the vehicle inspected by a professional technician to determine if your car will need an ECM replacement.

References

- Ludel, Moses (2011). "How-to: YJ Wrangler 4.2L Six Two-Barrel 'BBD' Carburetor Rebuild!". 4WD Mechanix Magazine. Retrieved 13 March 2014

- Engine-management-system: whyhighend.com, Retrieved 8 June, 2019

- Austen, Ian (2003-08-21). "WHAT'S NEXT; A Chip-Based Challenge to a Car's Spinning Camshaft". The New York Times. Retrieved 2009-01-16

- Symptoms-of-a-bad-or-failing-engine-control-module-ecm: yourmechanic.com, Retrieved 9 July, 2019

4

Engine Lubrication Systems

The engine lubrication system is used to distribute oil to the numerous moving parts of an engine for reducing friction between surfaces. Major engine lubrication systems include dry sump systems, wet sump system, total-loss oiling system and oil mist lubrication system. The topics elaborated in this chapter will help in gaining a better perspective about these engine lubrication systems.

Lubrication plays a key role in the life expectancy of an engine. Without oil, an engine would succumb to overheating and seizing very quickly. Lubricants help mitigate this problem, and if properly monitored and maintained, can extend the life of your motor.

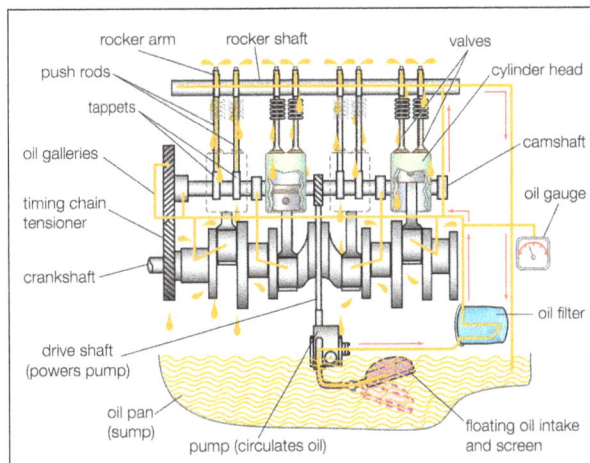

The engine lubrication system is to distribute oil to the moving parts to reduce friction between surfaces. Lubrication plays a key role in the life expectancy of an automotive engine. If the lubricating system fail, an engine would succumb to overheating and seizing very quickly. An oil pump is located on the bottom of the engine. The oil is pulled through a strainer, by the oil pump, removing larger contaminants from the mass of the fluid. The oil then forced through an oil filter under pressure to the main bearings and the oil pressure gauge. It is important to note that not all filters perform the same. A filter's ability to remove particles is dependent upon many factors, including the media material (pore size, surface area and depth of filter), the differential pressure across the media, and the flow rate across the media. From the main bearings, the oil passes into drilled passages in the crankshaft and the big-end bearings of the connecting rod. The oil fling dispersed by the rotating crankshaft lubricates the cylinder walls and piston-pin bearings. The excess oil is scraped off by the scraper rings on the piston. The engine oil also lubricates camshaft bearings and the timing chain or gears on the camshaft drive. The excess oil in the system then drains back to the sump.

Engine Oil

Superior quality engine oil is formulated with the high quality base oil and advanced technology based additive package to provide protection for automotive engines in severe service applications.

SAE Stands for the Society of Automotive Engineers, based in the U.S.A. The SAE grade specifies the most important parameters for engine oil mainly its viscosity. The SAE viscosity classification defines mainly viscosity limits at high and low temperature for any grade of lubricants. The SAE grade guide us to the right viscosity for different outside temperatures. Grades marked 'w' stand for winter are at a temperature below 0 °C.

API stands for the American Petroleum Institute. This body has specified the performance standards that oils used in road vehicles should meet. For oils to use in passenger car engines, the letters API are followed by a set of two letters such as SM, etc. Service Levels for passenger car oils or 'S' indicates for Spark Ignition Engine. These specified performance levels have evolved through the years, from API SA to SN.

Similarly, the API designates the performance of diesel engine oils with a letter sequence such as API CF-4.'C' indicates for commercial or compression ignition engine. Automotive gear oils they use API GL-4.API GL-5 etc.

The highest API for commercial engine oils (diesel oils) today is API CJ-4.

Purpose of Lubricating System

The lubricating oil performs the following functions in an engine:

- The oil lubricates moving parts to minimize wear by sealing the clearances between moving parts such as bearings, shafts, etc. Thus, the parts move on layers of oil, and not in direct contact with each other, which reduces power loss in the engine.

- The oil obtains heat from the moving parts of the engine which is transferred into the cooler oil in the oil pan. Therefore, the oil performs the function of a cooling agent. Some engines have oil nozzles which spray oil at the underside of the pistons, thereby removing heat from the pistons.

- The oil fills the clearances between rotating journals and the bearings. When heavy loads are abruptly placed on the bearings, the oil acts as a cushioning agent, which reduces the wear on bearings.

- The oil creates a seal between the walls of the cylinder and the piston rings, thereby reducing exhaust gas blowby.

- The oil performs the function of a cleaning agent by picking up dirt particles and taking them to oil pan. Larger particles are retained at the bottom while smaller particles are filtered out by oil filters.

Parts and Operation of Lubricating System

- Oil Pump: The gear-type oil pump has a pair of meshing gears. The spaces between the teeth are filled with oil when the gears unmesh. The oil pump obtains oil from the oil pan and sends oil through the oil filter to the oil galleries and main bearings. Some oil passes from the holes in the crankshaft to the rod bearings. Main bearings and rod bearings are lubricated adequately to achieve their desired objectives. In the rotor type oil pump, the inner rotor is driven and drives the outer rotor. As the rotor revolves, the gaps between the lobes are filled with oil. When the lobes of the inner rotor move into the gaps in the outer rotor, oil is forced out through the outlet of pump. An oil pump can also be driven by a camshaft gear that drives the ignition distributor or by the crankshaft.

- Oil Pan: Oil also flows to the cylinder head through drilled passages that make up the oil gallery, lubricates camshaft bearings and valves, and then returns to oil pan. Some engines have grooves or holes in connecting rods, which provide extra lubrication to pistons and walls of cylinders.

- Oil Cooler: Oil cooler prevents overheating of oil, by flow of engine coolant past tubes carrying hot oil. The coolant picks excess heat and carries it to the radiator.

- Oil Filter: The oil from oil pump flows through oil filter before reaching the engine bearings. The oil filter retains the dirt particles and allows only clean filtered oil to pass.

Lubricating System Indicators

There are indicator lights which are "on" when the engine oil pressure is low. Electric analog and electronic digital gauges are used to indicate the oil pressure. A dipstick is used to measure the oil level in the oil pan, while in some vehicles oil change indicator lights are used to identify the quality of oil.

Motor Oil

Adding motor oil.

Motor oil sample.

Motor oil, engine oil, or engine lubricant is any of various substances comprising base oils enhanced with particularly antiwear additive plus detergents, dispersants and, for multi-grade oils viscosity index improvers. Motor oil is used for lubrication of internal combustion engines. The main function of motor oil is to reduce friction and wear on moving parts and to clean the engine from sludge (one of the functions of dispersants) and varnish (detergents). It also neutralizes acids that originate from fuel and from oxidation of the lubricant (detergents), improves sealing of piston rings, and cools the engine by carrying heat away from moving parts.

In addition to the basic constituents noted in the preceding paragraph, almost all lubricating oils contain corrosion (GB: rust) and oxidation inhibitors. Motor oil may be composed of only a lubricant base stock in the case of non-detergent oil, or a lubricant base stock plus additives to improve the oil's detergency, extreme pressure performance, and ability to inhibit corrosion of engine parts.

Motor oils today are blended using base oils composed of petroleum-based hydrocarbons, polyalphaolefins (PAO) or their mixtures in various proportions, sometimes with up to 20% by weight of esters for better dissolution of additives.

Use

Motor oil is a lubricant used in internal combustion engines, which power cars, motorcycles, lawnmowers, engine-generators, and many other machines. In engines, there are parts which move against each other, and the friction wastes otherwise useful power by converting the kinetic energy to heat. It also wears away those parts, which could lead to lower efficiency and degradation of the engine. This increases fuel consumption and decreases power output and can lead to engine failure.

Lubricating oil creates a separating film between surfaces of adjacent moving parts to minimize direct contact between them, decreasing heat caused by friction and reducing wear, thus protecting the engine. In use, motor oil transfers heat through conduction as it flows through the engine. In an engine with a recirculating oil pump, this heat is transferred by means of airflow over the exterior surface of the [oil pan], airflow through an oil cooler and through oil gases evacuated by the Positive Crankcase Ventilation (PCV) system. While modern recirculating pumps are typically provided in passenger cars and other engines similar or larger in size, total loss oiling is a design option that remains popular in small and miniature engines.

In petrol (gasoline) engines, the top piston ring can expose the motor oil to temperatures of 160 °C (320 °F). In diesel engines, the top ring can expose the oil to temperatures over 315 °C (600 °F). Motor oils with higher viscosity indices thin less at these higher temperatures.

Coating metal parts with oil also keeps them from being exposed to oxygen, inhibiting oxidation at elevated operating temperatures preventing rust or corrosion. Corrosion inhibitors may also be added to the motor oil. Many motor oils also have detergents and dispersants added to help keep the engine clean and minimize oil sludge build-up. The oil is able to trap soot from combustion in itself, rather than leaving it deposited on the internal surfaces. It is a combination of this, and some singeing that turns used oil black after some running.

Rubbing of metal engine parts inevitably produces some microscopic metallic particles from the wearing of the surfaces. Such particles could circulate in the oil and grind against moving parts, causing wear. Because particles accumulate in the oil, it is typically circulated through an oil filter to remove harmful particles. An oil pump, a vane or gear pump powered by the engine, pumps the oil throughout the engine, including the oil filter. Oil filters can be a *full flow* or *bypass* type.

In the crankcase of a vehicle engine, motor oil lubricates rotating or sliding surfaces between the crankshaft journal bearings (main bearings and big-end bearings) and rods connecting the pistons to the crankshaft. The oil collects in an oil pan, or sump, at the bottom of the crankcase. In some small engines such as lawn mower engines, dippers on the bottoms of connecting rods dip into the oil at the bottom and splash it around the crankcase as needed to lubricate parts inside. In modern vehicle engines, the oil pump takes oil from the oil pan and sends it through the oil filter into oil galleries, from which the oil lubricates the main bearings holding the crankshaft up at the main journals and camshaft bearings operating the valves. In typical modern vehicles, oil pressure-fed from the oil galleries to the main bearings enters holes in the main journals of the crankshaft.

From these holes in the main journals, the oil moves through passageways inside the crankshaft to exit holes in the rod journals to lubricate the rod bearings and connecting rods. Some simpler designs relied on these rapidly moving parts to splash and lubricate the contacting surfaces between the piston rings and interior surfaces of the cylinders. However, in modern designs, there are also passageways through the rods which carry oil from the rod bearings to the rod-piston connections and lubricate the contacting surfaces between the piston rings and interior surfaces of the cylinders. This oil film also serves as a seal between the piston rings and cylinder walls to separate the combustion chamber in the cylinder head from the crankcase. The oil then drips back down into the oil pan.

Motor oil may also serve as a cooling agent. In some constructions, oil is sprayed through a nozzle inside the crankcase onto the piston to provide cooling of specific parts that undergo high-temperature strain. On the other hand, the thermal capacity of the oil pool has to be filled, i.e. the oil has to reach its designed temperature range before it can protect the engine under high load. This typically takes longer than heating the main cooling agent – water or mixtures thereof – up to its operating temperature. In order to inform the driver about the oil temperature, some older and most high-performance or racing engines feature an oil thermometer.

Non-Vehicle Motor Oils

An example is lubricating oil for four-stroke or four-cycle internal combustion engines such as those used in portable electricity generators and "walk behind" lawn mowers. Another example is two-stroke oil for lubrication of two-stroke or two-cycle internal combustion engines found in snow blowers, chain saws, model airplanes, gasoline-powered gardening equipment like hedge trimmers, leaf blowers and soil cultivators. Often, these motors are not exposed to as wide of service temperature ranges as in vehicles, so these oils may be single viscosity oils.

In small two-stroke engines, the oil may be pre-mixed with the gasoline or fuel, often in a rich gasoline:oil ratio of 25:1, 40:1 or 50:1, and burned in use along with the gasoline. Larger two-stroke engines used in boats and motorcycles may have a more economical oil injection system rather than oil pre-mixed into the gasoline. The oil injection system is not used on small engines used in applications like snowblowers and trolling motors as the oil injection system is too expensive for small engines and would take up too much room on the equipment. The oil properties will vary according to the individual needs of these devices. Non-smoking two-stroke oils are composed of esters or polyglycols. Environmental legislation for leisure marine applications, especially in Europe, encouraged the use of ester-based two cycle oil.

Properties

Most motor oils are made from a heavier, thicker petroleum hydrocarbon base stock derived from crude oil, with additives to improve certain properties. The bulk of a typical motor oil consists of hydrocarbons with between 18 and 34 carbon atoms per molecule. One of the most important properties of motor oil in maintaining a lubricating film between moving parts is its viscosity. The viscosity of a liquid can be thought of as its "thickness" or a measure of its resistance to flow. The viscosity must be high enough to maintain a lubricating film, but low enough that the oil can flow around the engine parts under all conditions. The viscosity index is a measure of how much the oil's viscosity changes as temperature changes. A higher viscosity index indicates the viscosity changes less with temperature than a lower viscosity index.

Motor oil must be able to flow adequately at the lowest temperature it is expected to experience in order to minimize metal to metal contact between moving parts upon starting up the engine. The *pour point* defined first this property of motor oil, as defined by ASTM D97 as "an index of the lowest temperature of its utility" for a given application, but the cold-cranking simulator (CCS,) and mini-rotary viscometer (MRV, ASTM D4684-08) are today the properties required in motor oil specs and define the SAE classifications.

Oil is largely composed of hydrocarbons which can burn if ignited. Still another important property of motor oil is its flash point, the lowest temperature at which the oil gives off vapors which can ignite. It is dangerous for the oil in a motor to ignite and burn, so a high flash point is desirable. At a petroleum refinery, fractional distillation separates a motor oil fraction from other crude oil fractions, removing the more volatile components, and therefore increasing the oil's flash point (reducing its tendency to burn).

Another manipulated property of motor oil is its total base number (TBN), which is a measurement of the reserve alkalinity of an oil, meaning its ability to neutralize acids. The resulting quantity is determined as mg KOH/ (gram of lubricant). Analogously, total acid number (TAN) is the measure of a lubricant's acidity. Other tests include zinc, phosphorus, or sulfur content, and testing for excessive foaming.

The Noack volatility test determines the physical evaporation loss of lubricants in high temperature service. A maximum of 14% evaporation loss is allowable to meet API SL and ILSAC GF-3 specifications. Some automotive OEM oil specifications require lower than 10%.

Viscosity Grades

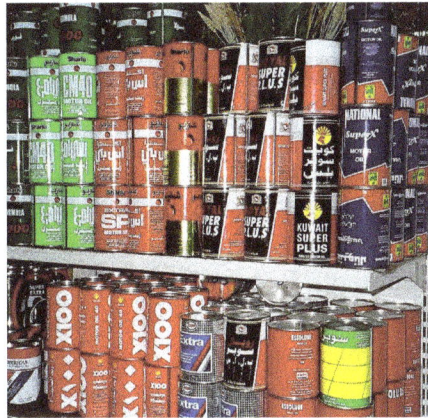

Range of motor oils on display in Kuwait in obsolete cardboard cans with steel lids.

The Society of Automotive Engineers (SAE) has established a numerical code system for grading motor oils according to their viscosity characteristics. The original viscosity grades were all mono-grades, e.g. a typical engine oil was a SAE 30. This is because as all oils thin when heated, so to get the right film thickness at operating temperatures oil manufacturers needed to start with a thick oil. This meant that in cold weather it would be difficult to start the engine as the oil was too thick to crank. However, oil additive technology was introduced that allowed oils to thin more slowly (i.e. to retain a higher viscosity index); this allowed selection of a thinner oil to start with, e.g. "SAE 15W-30", a product that acts like an SAE 15 at cold temperatures (15W for winter) and like an SAE 30 at 100 °C (212 °F).

Therefore, there is one set which measures cold temperature performance (0W, 5W, 10W, 15W and 20W). The second set of measurements is for high temperature performance (8, 12, 16, 20, 30, 40, 50). The document SAE J300 defines the viscometrics related to these grades.

Kinematic viscosity is graded by measuring the time it takes for a standard amount of oil to flow through a standard orifice at standard temperatures. The longer it takes, the higher the viscosity and thus the higher the SAE code. Larger numbers are thicker.

The SAE has a separate viscosity rating system for gear, axle, and manual transmission oils, SAE J306, which should not be confused with engine oil viscosity. The higher numbers of a gear oil (e.g., 75W-140) do not mean that it has higher viscosity than an engine oil. In anticipation of new lower engine oil viscosity grades, to avoid confusion with the "winter" grades of oil the SAE adopted SAE 16 as a standard to follow SAE 20 instead of SAE 15. Regarding the change Michael Covitch of Lubrizol, Chair of the SAE International Engine Oil Viscosity Classification (EOVC) task force was quoted stating "If we continued to count down from SAE 20 to 15 to 10, etc., we would be facing continuing customer confusion problems with popular low-temperature viscosity grades such as SAE 10W, SAE 5W, and SAE 0W," he noted. "By choosing to call the new viscosity grade SAE 16, we established a precedent for future grades, counting down by fours instead of fives: SAE 12, SAE 8, SAE 4."

Single-grade

A single-grade engine oil, as defined by SAE J300, cannot use a polymeric viscosity index improver (VII, also viscosity modifier, VM) additive. SAE J300 has established eleven viscosity grades, of which six are considered Winter-grades and given a W designation. The 11 viscosity grades are 0W, 5W, 10W, 15W, 20W, 25W, 20, 30, 40, 50, and 60. These numbers are often referred to as the "weight" of a motor oil, and single-grade motor oils are often called "straight-weight" oils.

For single winter grade oils, the dynamic viscosity is measured at different cold temperatures, specified in J300 depending on the viscosity grade, in units of mPa·s, or the equivalent older non-SI units, centipoise (abbreviated cP), using two different test methods. They are the cold-cranking simulator and the mini-rotary viscometer. Based on the coldest temperature the oil passes at, that oil is graded as SAE viscosity grade 0W, 5W, 10W, 15W, 20W, or 25W. The lower the viscosity grade, the lower the temperature the oil can pass. For example, if an oil passes at the specifications for 10W and 5W, but fails for 0W, then that oil must be labeled as an SAE 5W. That oil cannot be labeled as either 0W or 10W.

For single non-winter grade oils, the kinematic viscosity is measured at a temperature of 100 °C (212 °F) in units of mm²/s (millimeter squared per second) or the equivalent older non-SI units, centistokes (abbreviated cSt). Based on the range of viscosity the oil falls in at that temperature, the oil is graded as SAE viscosity grade 20, 30, 40, 50, or 60. In addition, for SAE grades 20, 30, and 40, a minimum viscosity measured at 150 °C (302 °F) and at a high-shear rate is also required. The higher the viscosity, the higher the SAE viscosity grade is.

Multi-grade

The temperature range the oil is exposed to in most vehicles can be wide, ranging from cold temperatures in the winter before the vehicle is started up to hot operating temperatures when the vehicle is fully warmed up in hot summer weather. A specific oil will have high viscosity when cold and a lower viscosity at the engine's operating temperature. The difference in viscosities for most single-grade oil is too large between the extremes of temperature. To bring the difference in viscosities closer together, special polymer additives called viscosity index improvers, or VIIs, are added to the oil. These additives are used to make the oil a *multi-grade* motor oil, though it is possible to have a multi-grade oil without the use of VIIs. The idea is to cause the multi-grade oil to have the viscosity of the base grade when cold and the viscosity of the second grade when hot. This enables one type of oil to be used all year. In fact, when multi-grades were initially developed, they were frequently described as *all-season oil*. The viscosity of a multi-grade oil still varies logarithmically with temperature, but the slope representing the change is lessened.

The SAE designation for multi-grade oils includes two viscosity grades; for example, *10W-30* designates a common multi-grade oil. The first number '10W' is the equivalent grade of the single grade oil that has the oil's viscosity at cold temperature and the second number is the grade of the equivalent single-grade oil that describes its viscosity at 100 °C (212 °F). Note that both numbers are grades and not viscosity values. The two numbers used are individually defined by SAE J300 for single-grade oils. Therefore, an oil labeled as 10W-30 must pass the SAE J300 viscosity grade requirement for both 10W and 30, and all limitations placed on the viscosity grades (for example, a 10W-30 oil must fail the J300 requirements at 5W). Also, if an oil does not contain any VIIs, and can pass as a multi-grade, that oil can be labeled with either of the two SAE viscosity grades. For example, a very simple multi-grade oil that can be easily made with modern base oils without any VII is a 20W-20. This oil can be labeled as 20W-20, 20W, or 20. Note, if any VIIs are used, however, then that oil cannot be labeled as a single grade.

Breakdown of VIIs under shear is a concern in motorcycle applications, where the transmission may share lubricating oil with the motor. For this reason, motorcycle-specific oil is sometimes recommended. The necessity of higher-priced motorcycle-specific oil has also been challenged by at least one consumer organization.

Standards

American Petroleum Institute (API)

Engine lubricants are evaluated against the American Petroleum Institute (API), SJ, SL, SM, SN, CH-4, CI-4, CI-4 PLUS, CJ-4, CK and FA as well as International Lubricant Standardization and Approval Committee (ILSAC) GF-3, GF-4 and GF-5, and Cummins, Mack and John Deere (and other Original Equipment Manufacturers (OEM)) requirements. These evaluations include chemical and physical properties using bench test methods as well as actual running engine tests to quantify engine sludge, oxidation, component wear, oil consumption, piston deposits and fuel economy.

The API sets minimum performance standards for lubricants. Motor oil is used for the lubrication, cooling, and cleaning of internal combustion engines. Motor oil may be composed of only a lubricant base stock in the case of mostly obsolete non-detergent oil, or a lubricant base stock plus additives to improve the oil's detergency, extreme pressure performance, and ability to inhibit corrosion of engine parts.

Groups: Lubricant base stocks are categorized into five groups by the API. Group I base stocks are composed of fractionally distilled petroleum which is further refined with solvent extraction processes to improve certain properties such as oxidation resistance and to remove wax. Poorly refined mineral oils that fail to meet the minimum VI of 80 required in group I fit into Group V. Group II base stocks are composed of fractionally distilled petroleum that has been hydrocracked to further refine and purify it. Group III base stocks have similar characteristics to Group II base stocks, except that Group III base stocks have higher viscosity indexes. Group III base stocks are produced by further hydrocracking of either Group II base stocks or hydroisomerized slack wax (a Group I and II dewaxing process by-product). Group IV base stock are polyalphaolefins (PAOs). Group V is a catch-all group for any base stock not described by Groups I to IV. Examples of group

V base stocks include polyolesters (POE), polyalkylene glycols (PAG), and perfluoropolyalkylethers (PFPAEs) and poorly refined mineral oil. Groups I and II are commonly referred to as mineral oils, group III is typically referred to as synthetic (except in Germany and Japan, where they must not be called synthetic) and group IV is a synthetic oil. Group V base oils are so diverse that there is no catch-all description.

The API service classes have two general classifications: *S* for "service/spark ignition" (typical passenger cars and light trucks using gasoline engines), and *C* for "commercial/compression ignition" (typical diesel equipment). Engine oil which has been tested and meets the API standards may display the API Service Symbol (also known as the "Donut") with the service categories on containers sold to oil users.

The latest API service category is API SN for gasoline automobile and light-truck engines. The SN standard refers to a group of laboratory and engine tests, including the latest series for control of high-temperature deposits. Current API service categories include SN, SM, SL and SJ for gasoline engines. All previous service categories are obsolete, although motorcycle oils commonly still use the SF/SG standard.

All the current gasoline categories (including the obsolete SH) have placed limitations on the phosphorus content for certain SAE viscosity grades (the xW-20, xW-30) due to the chemical poisoning that phosphorus has on catalytic converters. Phosphorus is a key anti-wear component in motor oil and is usually found in motor oil in the form of zinc dithiophosphate (ZDDP). Each new API category has placed successively lower phosphorus and zinc limits, and thus has created a controversial issue of obsolescent oils needed for older engines, especially engines with sliding (flat/cleave) tappets. API and ILSAC, which represents most of the world's major automobile/engine manufacturers, state API SM/ILSAC GF-4 is fully backwards compatible, and it is noted that one of the engine tests required for API SM, the Sequence IVA, is a sliding tappet design to test specifically for cam wear protection. Not everyone is in agreement with backwards compatibility, and in addition, there are special situations, such as "performance" engines or fully race built engines, where the engine protection requirements are above and beyond API/ILSAC requirements. Because of this, there are specialty oils out in the market place with higher than API allowed phosphorus levels. Most engines built before 1985 have the flat/cleave bearing style systems of construction, which is sensitive to reducing zinc and phosphorus. For example, in API SG rated oils, this was at the 1200–1300 ppm level for zinc and phosphorus, where the current SM is under 600 ppm. This reduction in anti-wear chemicals in oil has caused premature failures of camshafts and other high pressure bearings in many older automobiles and has been blamed for premature failure of the oil pump drive/cam position sensor gear that is meshed with camshaft gear in some modern engines.

The current diesel engine service categories are API CK-4, CJ-4, CI-4 PLUS, CI-4, CH-4, and FA-4. The previous service categories such as API CC or CD are obsolete. API solved problems with API CI-4 by creating a separate API CI-4 PLUS category that contains some additional requirements – this marking is located in the lower portion of the API Service Symbol "Donut".

API CK-4 and FA-4 have been introduced for 2017 model American engines. API CK-4 is backward compatible that means API CK-4 oils are assumed to provide superior performance to oils made to previous categories and could be used without problems in all previous model engines.

API FA-4 oils are different (that is why API decided to start a new group in addition to API Sx and API Cx). API FA-4 oils are formulated for enhanced fuel economy (presented as reduced greenhouse gas emission). To achieve that, they are SAE xW-30 oils blended to a high temperature high shear viscosity from 2.9 cP to 3.2 cP. They are not suitable for all engines thus their use depends on the decision of each engine manufacturer. They cannot be used with diesel fuel containing more than 15 ppm sulfur.

Cummins reacted to the introduction of API CK-4 and API FA-4 by issuing its CES 20086 list of API CK-4 registered oils and CES 20087 list of API FA-4 registered oils. Valvoline oils are preferred. Ford does not recommend API CK-4 or FA-4 oils in its diesel engines.

While engine oils are formulated to meet a specific API service category, they in fact conform closely enough to both the gasoline and diesel categories. Thus diesel rated engine oils usually carry the relevant gasoline categories, e.g. an API CJ-4 oil could show either API SL or API SM on the container. The rule is that the first mentioned category is fully met and the second one is fully met except where its requirements clash with the requirements of the first one.

Motorcycle Oil

The API oil classification structure has eliminated specific support for wet-clutch motorcycle applications in their descriptors, and API SJ and newer oils are referred to be specific to automobile and light truck use. Accordingly, motorcycle oils are subject to their own unique standards, motorcycle oils commonly still use the obsolescent SF/SG standard.

ILSAC

The International Lubricant Standardization and Approval Committee (ILSAC) also has standards for motor oil. Introduced in 2004, GF-4 applies to SAE 0W-20, 5W-20, 0W-30, 5W-30, and 10W-30 viscosity grade oils. In general, ILSAC works with API in creating the newest gasoline oil specification, with ILSAC adding an extra requirement of fuel economy testing to their specification. For GF-4, a Sequence VIB Fuel Economy Test is required that is not required in API service category SM.

A key new test for GF-4, which is also required for API SM, is the Sequence IIIG, which involves running a 3.8 litres (230 cu in), GM 3.8 L V-6 at 125 hp (93 kW), 3,600 rpm, and 150 °C (302 °F) oil temperature for 100 hours. These are much more severe conditions than any API-specified oil was designed for: cars which typically push their oil temperature consistently above 100 °C (212 °F) are most turbocharged engines, along with most engines of European or Japanese origin, particularly small capacity, high power output.

The IIIG test is about 50% more difficult than the previous IIIF test, used in GF-3 and API SL oils. Engine oils bearing the API starburst symbol since 2005 are ILSAC GF-4 compliant.

To help consumers recognize that an oil meets the ILSAC requirements, API developed a "starburst" certification mark.

A new set of specifications, GF-5, took effect in October 2010. The industry had one year to convert their oils to GF-5 and in September 2011, ILSAC no longer offered licensing for GF-4.

ACEA

The ACEA (Association des Constructeurs Européens d'Automobiles) performance/quality classifications A3/A5 tests used in Europe are arguably more stringent than the API and ILSAC standards. CEC (The Co-ordinating European Council) is the development body for fuel and lubricant testing in Europe and beyond, setting the standards via their European Industry groups; ACEA, ATIEL, ATC and CONCAWE.

Lubrizol, a supplier of additives to nearly all motor oil companies, hosts a Relative Performance Tool which directly compares the manufacturer and industry specs. Differences in their performance is apparent in the form of interactive spider graphs, which both expert and novice can appreciate.

ACEA A5/B5 Oil grade refers to synthetic oils.

JASO

The Japanese Automotive Standards Organization (JASO) has created their own set of performance and quality standards for petrol engines of Japanese origin.

For four-stroke gasoline engines, the JASO T904 standard is used, and is particularly relevant to motorcycle engines. The JASO T904-MA and MA2 standards are designed to distinguish oils that are approved for wet clutch use, with MA2 lubricants delivering higher friction performance. The JASO T904-MB standard denotes oils not suitable for wet clutch use, and are therefore used in scooters equipped with continuously variable transmissions. The addition of friction modifiers to JASO MB oils can contribute to greater fuel economy in these applications.

For two-stroke gasoline engines, the JASO M345 (FA, FB, FC, FD) standard is used, and this refers particularly to low ash, lubricity, detergency, low smoke and exhaust blocking.

These standards, especially JASO-MA (for motorcycles) and JASO-FC, are designed to address oil-requirement issues not addressed by the API service categories. One element of the JASO-MA standard is a friction test designed to determine suitability for wet clutch usage. An oil that meets JASO-MA is considered appropriate for wet clutch operations. Oils marketed as motorcycle-specific will carry the JASO-MA label.

ASTM

A 1989 American Society for Testing and Materials (ASTM) report stated that its 12-year effort to come up with a new high-temperature, high-shear (HTHS) standard was not successful. Referring to SAE J300, the basis for current grading standards, the report stated:

> "The rapid growth of non-Newtonian multigraded oils has rendered kinematic viscosity as a nearly useless parameter for characterising "real" viscosity in critical zones of an engine. There are those who are disappointed that the twelve-year effort has not resulted in a redefinition of the SAE J300 Engine Oil Viscosity Classification document so as to express high-temperature viscosity of the various grades. In the view of this writer, this redefinition did not occur because the automotive lubricant market knows of no field failures unambiguously attributable to insufficient HTHS oil viscosity".

Other Additives

In addition to the viscosity index improvers, motor oil manufacturers often include other additives such as detergents and dispersants to help keep the engine clean by minimizing sludge buildup, corrosion inhibitors, and alkaline additives to neutralize acidic oxidation products of the oil. Most commercial oils have a minimal amount of zinc dialkyldithiophosphate as an anti-wear additive to protect contacting metal surfaces with zinc and other compounds in case of metal to metal contact. The quantity of zinc dialkyldithiophosphate is limited to minimize adverse effect on catalytic converters. Another aspect for after-treatment devices is the deposition of oil ash, which increases the exhaust back pressure and reduces fuel economy over time. The so-called "chemical box" limits today the concentrations of sulfur, ash and phosphorus (SAP).

There are other additives available commercially which can be added to the oil by the user for purported additional benefit. Some of these additives include:

- Antiwear additives, like zinc dialkyldithiophosphate (ZDDP) and its alternatives due to phosphorus limits in some specifications. Calcium sulfonates additives are also added to protect motor oil from oxidative breakdown and to prevent the formation of sludge and varnish deposits. Both were the main basis of additive packages used by lubricant manufacturers up until the 1990s when the need for ashless additives arose. Main advantage was very low price and wide availability (sulfonates were originally waste byproducts). Currently there are ashless oil lubricants without these additives, which can only fulfill the qualities of the previous generation with more expensive basestock and more expensive organic or organometallic additive compounds. Some new oils are not formulated to provide the level of protection of previous generations to save manufacturing costs.

- Some molybdenum disulfide containing additives to lubricating oils are claimed to reduce friction, bond to metal, or have anti-wear properties. MoS_2 particles can be shear-welded on steel surface and some engine components were even treated with MoS_2 layer during manufacture, namely liners in engines. (Trabant for example). They were used in World War II in flight engines and became commercial after World War II until the 1990s. They were commercialized in the 1970s (ELF ANTAR Molygraphite) and are today still available (Liqui Moly MoS_2 10 W-40). Main disadvantage of molybdenum disulfide is anthracite black color, so oil treated with it is hard to distinguish from a soot filled engine oil with metal shavings from spun crankshaft bearing.

- In the 1980s and 1990s, additives with suspended PTFE particles were available, e.g., "Slick50," to consumers to increase motor oil's ability to coat and protect metal surfaces. There is controversy as to the actual effectiveness of these products, as they can coagulate and clog the oil filter and tiny oil passages in the engine. It is supposed to work under boundary lubricating conditions, which good engine designs tend to avoid anyway. Also, Teflon alone has little to no ability to firmly stick on a sheared surface, unlike molybdenum disulfide.

- Many patents proposed use perfluoropolymers to reduce friction between metal parts, such as PTFE (Teflon), or micronized PTFE. However, the application obstacle of PTFE is insolubility in lubricant oils. Their application is questionable and depends mainly on the engine design – one that can not maintain reasonable lubricating conditions might benefit, while properly designed engine with oil film thick enough would not see any difference. PTFE is a very soft material, thus its friction coefficient becomes worse than that of hardened steel-to-steel mating surfaces under common loads. PTFE is used in composition of sliding bearings where it improves lubrication under relatively light load until the oil pressure builds up to full hydrodynamic lubricating conditions.

Some molybdenum disulfide containing oils may be unsuitable for motorcycles which share wet clutch lubrication with the engine.

Environmental Effects

Blue drain and yellow fish symbol used by the UK Environment Agency to raise awareness of the ecological impacts of contaminating surface drainage.

Due to its chemical composition, worldwide dispersion and effects on the environment, used motor oil is considered a serious environmental problem. Most current motor oil lubricants contain petroleum base stocks, which are toxic to the environment and difficult to dispose of after use. Over 40% of the pollution in America's waterways is from used motor oil. Used oil is considered the largest source of oil pollution in the U.S. harbor and waterways, at 1,460 ML (385×10^6 US gal) per year, mostly from improper disposal. By far, the greatest cause of motor oil pollution in our oceans comes from drains and urban street runoff, much of which is from improper disposal of engine oil. One US gallon (3.8 l) of used oil can create a 32,000 m² (8 acres) slick on surface water, threatening fish, waterfowl and other aquatic life. According to the U.S. EPA, films of oil on the surface of water prevent the replenishment of dissolved oxygen, impair photosynthetic processes, and block sunlight. Toxic effects of used oil on freshwater and marine organisms vary, but significant long-term effects have been found at concentrations of 310 ppm in several freshwater fish species and as low as 1 ppm in marine life forms. Motor oil can have an incredibly detrimental effect on the environment, particularly to plants that depend on healthy soil to grow. There are three main ways that motor oil affects plants: contaminating water supplies, contaminating soil, and poisoning plants. Used motor oil dumped on land reduces soil productivity. Improperly disposed used oil ends up in landfills, sewers, backyards, or storm drains where soil, groundwater and drinking water may be contaminated.

Synthetic Oils

Synthetic lubricants were first synthesized, or man-made, in significant quantities as replacements for mineral lubricants (and fuels) by German scientists in the late 1930s and early 1940s because of their lack of sufficient quantities of crude for their (primarily military) needs. A significant factor in its gain in popularity was the ability of synthetic-based lubricants to remain fluid in the sub-zero temperatures of the Eastern front in wintertime, temperatures which caused petroleum-based lubricants to solidify owing to their higher wax content. The use of synthetic lubricants widened through the 1950s and 1960s owing to a property at the other end of the temperature spectrum – the ability to lubricate aviation engines at high temperatures that caused mineral-based lubricants to break down. In the mid-1970s, synthetic motor oils were formulated and commercially applied for the first time in automotive applications. The same SAE system for designating motor oil viscosity also applies to synthetic oils.

Synthetic oils are derived from either Group III, Group IV, or some Group V bases. Synthetics include classes of lubricants like synthetic esters (Group V) as well as "others" like GTL (methane gas-to-liquid) (Group III +) and polyalpha-olefins (Group IV). Higher purity and therefore better property control theoretically means synthetic oil has better mechanical properties at extremes of high and low temperatures. The molecules are made large and "soft" enough to retain good viscosity at higher temperatures, yet branched molecular structures interfere with solidification and therefore allow flow at lower temperatures. Thus, although the viscosity still decreases as temperature increases, these synthetic motor oils have a higher viscosity index over the traditional petroleum base. Their specially designed properties allow a wider temperature range at higher and lower temperatures and often include a lower pour point. With their improved viscosity index, synthetic oils need lower levels of viscosity index improvers, which are the oil components most vulnerable to thermal and mechanical degradation as the oil ages, and thus they do not degrade as quickly as traditional motor oils. However, they still fill up with particulate matter, although the matter better suspends within the oil, and the oil filter still fills and clogs up over time. So periodic oil and filter changes should still be done with synthetic oil, but some synthetic oil suppliers suggest that the intervals between oil changes can be longer, sometimes as long as 16,000–24,000 kilometres (9,900–14,900 mi) primarily due to reduced degradation by oxidation.

Tests show that fully synthetic oil is superior in extreme service conditions to conventional oil, and may perform better for longer under standard conditions. But in the vast majority of vehicle applications, mineral oil-based lubricants, fortified with additives and with the benefit of over a century of development, continue to be the predominant lubricant for most internal combustion engine applications.

Bio-based Oils

Bio-based oils existed prior to the development of petroleum-based oils in the 19th century. They have become the subject of renewed interest with the advent of bio-fuels and the push for green products. The development of canola-based motor oils began in 1996 in order to pursue environmentally friendly products. Purdue University has funded a project to develop and test such oils. Test results indicate satisfactory performance from the oils tested. A review on the status of bio-based motor oils and base oils globally, as well as in the U.S, shows how bio-based lubricants show promise in augmenting the current petroleum-based supply of lubricating materials, as well as replacing it in many cases.

The USDA National Center for Agricultural Utilization Research developed an Estolide lubricant technology made from vegetable and animal oils. Estolides have shown great promise in a wide range of applications, including engine lubricants. Working with the USDA, a California-based company Biosynthetic Technologies has developed a high-performance "drop-in" biosynthetic oil using Estolide technology for use in motor oils and industrial lubricants. This biosynthetic oil American Petroleum Institute (API) has the potential to greatly reduce environmental challenges associated with petroleum. Independent testing not only shows biosynthetic oils to be among the highest-rated products for protecting engines and machinery; they are also bio-based, biodegradable, non-toxic and do not bioaccumulate in marine organisms. Also, motor oils and lubricants formulated with biosynthetic base oils can be recycled and re-refined with petroleum-based oils. The U.S.-based company Green Earth Technologies manufactures a bio-based motor oil, called G-Oil, made from animal oils.

Maintenance

Checking oil level (Togo).

Oil being drained from a car.

The oil and the oil filter need to be periodically replaced. While there is a full industry surrounding regular oil changes and maintenance, an oil change is a fairly simple operation that most car owners can do themselves.

In engines, there is some exposure of the oil to products of internal combustion, and microscopic coke particles from black soot accumulate in the oil during operation. Also, the rubbing of metal engine parts produces some microscopic metallic particles from the wearing of the surfaces. Such particles could circulate in the oil and grind against the part surfaces causing wear. The oil filter removes many of the particles and sludge, but eventually, the oil filter can become clogged, if used for extremely long periods.

The motor oil and especially the additives also undergo thermal and mechanical degradation, which reduce the viscosity and reserve alkalinity of the oil. At reduced viscosity, the oil is not as capable of lubricating the engine, thus increasing wear and the chance of overheating. Reserve alkalinity is the ability of the oil to resist the formation of acids. Should the reserve alkalinity decline to zero, those acids form and corrode the engine.

Some engine manufacturers specify which SAE viscosity grade of oil should be used, but different viscosity motor oil may perform better based on the operating environment. Many manufacturers have varying requirements and have designations for motor oil they require to be used. This is driven by the EPA requirement that the same viscosity grade of oil used in the MPG test must be recommended to the customer. This exclusive recommendation lead to the elimination of informative charts depicting climate temp range along with several corresponding oil viscosity grades being suggested.

Oil change at oil change shop.

In general, unless specified by the manufacturer, thicker oils are not necessarily better than thinner oils; heavy oils tend to stick longer to parts between two moving surfaces, and this degrades the oil faster than a lighter oil that flows better, allowing fresh oil in its place sooner. Cold weather has a thickening effect on conventional oil, and this is one reason thinner oils are manufacturer recommended in places with cold winters.

Motor oil changes are usually scheduled based on the time in service or the distance that the vehicle has traveled. These are rough indications of the real factors that control when an oil change is appropriate, which include how long the oil has been run at elevated temperatures, how many heating cycles the engine has been through, and how hard the engine has worked. The vehicle distance is intended to estimate the time at high temperature, while the time in service is supposed to correlate with the number of vehicle trips and capture the number of heating cycles. Oil does not degrade significantly just sitting in a cold engine. On the other hand, if a car is driven just for very short distances, the oil will not fully heat up, and it will accumulate contaminants such as water, due to lack of sufficient heat to boil off the water. Oil in this condition, just sitting in an engine, can cause problems.

Also important is the quality of the oil used, especially with synthetics (synthetics are more stable than conventional oils). Some manufacturers address this (for example, BMW and VW with their respective long-life standards), while others do not.

Time-based intervals account for the short-trip drivers who drive short distances, which build up more contaminants. Manufacturers advise to not exceed their time or distance-driven interval for a motor oil change. Many modern cars now list somewhat higher intervals for changing oil and filter, with the constraint of "severe" service requiring more frequent changes with less-than-ideal driving. This applies to short trips of under 15 kilometres (10 mi), where the oil does not get to full operating temperature long enough to burn off condensation, excess fuel, and other contamination that leads to "sludge", "varnish", "acids", or other deposits. Many manufacturers have engine computer calculations to estimate the oil's condition based on the factors which degrade it, such as RPM, temperatures, and trip length; one system adds an optical sensor for determining the clarity of the oil in the engine. These systems are commonly known as Oil Life Monitors or OLMs.

Some quick oil change shops recommended intervals of 5,000 kilometres (3,000 mi) or every three months, which is not necessary, according to many automobile manufacturers. This has led to a campaign by the California EPA against the 3,000 mile myth, promoting vehicle manufacturer's recommendations for oil change intervals over those of the oil change industry.

The engine user can, in replacing the oil, adjust the viscosity for the ambient temperature change, thicker for summer heat and thinner for the winter cold. Lower-viscosity oils are common in newer vehicles.

By the mid-1980s, recommended viscosities had moved down to 5W-30, primarily to improve fuel efficiency. A typical modern application would be Honda motor's use of 5W-20 (and in their newest vehicles, 0W-20) viscosity oil for 12,000 kilometres (7,500 mi). Engine designs are evolving to allow the use of even lower-viscosity oils without the risk of excessive metal-to-metal abrasion, principally in the cam and valve mechanism areas. In line with car manufacturers push towards these lower viscosities in search of better fuel economy, on April 2, 2013 the Society of Automotive Engineers (SAE) introduced an SAE 16 viscosity rating, a break from its traditional "divisible by 10" numbering system for its high-temperature viscosity ratings that spanned from low-viscosity SAE 20 to high-viscosity SAE 60.

Future

A new process to break down polyethylene, a common plastic product found in many consumer containers, is used to make a paraffin-like wax with the correct molecular properties for conversion into a lubricant, bypassing the expensive Fischer–Tropsch process. The plastic is melted and then pumped into a furnace. The heat of the furnace breaks down the molecular chains of polyethylene into wax. Finally, the wax is subjected to a catalytic process that alters the wax's molecular structure, leaving a clear oil.

Biodegradable Motor Oils based on esters or hydrocarbon-ester blends appeared in the 1990s followed by formulations beginning in 2000 which respond to the bio-no-tox-criteria of the European preparations directive. This means, that they not only are biodegradable according to OECD 301x test methods, but also the aquatic toxicities (fish, algae, daphnie) are each above 100 mg/L.

Another class of base oils suited for engine oil are the polyalkylene glycols. They offer zero-ash, bio-no-tox properties and lean burn characteristics.

Re-refined Motor Oil

The oil in a motor oil product does break down and burns as it is used in an engine – it also gets contaminated with particles and chemicals that make it a less effective lubricant. Re-refining cleans the contaminants and used additives out of the dirty oil. From there, this clean "base stock" is blended with some virgin base stock and a new additives package to make a finished lubricant product that can be just as effective as lubricants made with all-virgin oil. The United States Environmental Protection Agency (EPA) defines re-refined products as containing at least 25% re-refined base stock, but other standards are significantly higher. The California State public contract code defines a re-refined motor oil as one that contains at least 70% re-refined base stock.

Packaging

Motor oils were sold at retail in glass bottles, metal cans and metal/cardboard cans, before the advent of the current polyethylene plastic bottle, which began to appear in the early 1980s. Reusable spouts were made separately from the cans; with a piercing point like that of a can opener, these spouts could be used to puncture the top of the can and to provide an easy way to pour the oil.

Today, motor oil in the USA is generally sold in bottles of 1 US quart (950 mL) and on a rarity in 1-litre (34 US fl oz) as well as in larger plastic containers ranging from approximately 4.4 to 5 liters (4.6 to 5.3 U.S. qt) due to most small to mid-size engines requiring around 3.6 to 5.2 liters (3.8 to 5.5 U.S. qt) of engine oil. In the rest of the world, it is most commonly available in 1L, 3L, 4L and 5L retail packages.

There was a growing trend to sell motor oil in flexible packaging, for instance in stand-up pouches. However, as motor oil pouch packs are made with a multi-layered plastic laminate of nylon, polyester and LLDPE which is difficult to recycle, the growth in flexible motor oil packaging may be limited.

Distribution to larger users (such as drive-through oil change shops) is often in bulk, by tanker truck or in 1 barrel (160 l) drums.

Two-stroke Oil

An example of two-stroke oil bottle with measurement cap. Oil is dyed blue to make it easier to recognize it in the gasoline. Because it's not diluted, it appears black in this bottle.

Two-stroke oil (also referred to as two-cycle oil, 2-cycle oil, 2T oil, or 2-stroke oil) is a special type of motor oil intended for use in crankcase compression two-stroke engines. Unlike a four-stroke engine, whose crankcase is closed except for its ventilation system, a two-stroke engine uses the crankcase as part of the induction tract, and therefore, oil must be mixed with gasoline to be distributed throughout the engine for lubrication. The resultant mix is referred to as premix or petroil. This oil is ultimately burned along with the fuel as a total-loss oiling system. This results in increased exhaust emissions, sometimes with excess smoke and/or a distinctive odor.

The oil-base stock can be petroleum, castor oil, semi-synthetic or synthetic oil and is mixed (or metered by injection) with petrol/gasoline at a volumetric fuel-to-oil ratio ranging from 16:1 to as low as 100:1. To avoid the high emissions and oily deposits on spark plugs, modern two-strokes, especially for small engines such as garden equipment and chainsaws, may now demand a synthetic oil and can suffer from oiling problems otherwise.

Engine original equipment manufacturers (OEMs) introduced pre-injection systems (sometimes known as "auto-lube") to engines to operate from a 32:1 to 100:1 ratio. Oils must meet or exceed the following typical specifications: TC-W3TM, NMMA, [API] TC, JASO FC, ISO-L-EGC.

Comparing regular lubricating oil with two-stroke oil, the relevant difference is that two-stroke oil must have a much lower ash content. This is required to minimize deposits that tend to form if ash is present in the oil which is burned in the engine's combustion chamber. Additionally a non-2T-specific oil can turn to gum in a matter of days if mixed with gasoline and not immediately consumed. Another important factor is that 4-stroke engines have a different requirement for 'stickiness' than 2-strokes do. Since the 1980s different types of two-stroke oil have been developed for specialized uses such as outboard motor two-strokes, premix two-stroke oil, as well as the more standard auto lube (motorcycle) two-stroke oil. As a rule of thumb, most containers of oil commercially offered will have somewhere on the label printed that it is compatible with 'Autolube' or injector pumps. Those bottles tend to have the consistency of liquid dish soap if shaken. A more viscous oil cannot reliably be passed through an injection system, although a premix machine can be run on either type.

"Racing" oil or castor-based does offer excellent lubricity - at the expense of premature coking. For the average moped/scooter/trail rider it will not garner an appreciable increase in performance and will require very frequent teardowns.

Additive Ingredients

Additives for two-stroke oils fall into several general categories: Detergent/Dispersants, Antiwear agents, Biodegradability components and antioxidants (Zinc compounds). Some of the higher quality include a fuel stabilizer as well.

Dry Sump

A dry-sump system is a method to manage the lubricating motor oil in four-stroke and large two-stroke piston driven internal combustion engines. The dry-sump system uses two or more oil pumps and a separate oil reservoir, as opposed to a conventional wet-sump system, which uses only the main sump (U.S.: oil pan) below the engine and a single pump. A dry-sump engine requires a pressure relief valve to regulate negative pressure inside the engine, so internal seals are not inverted.

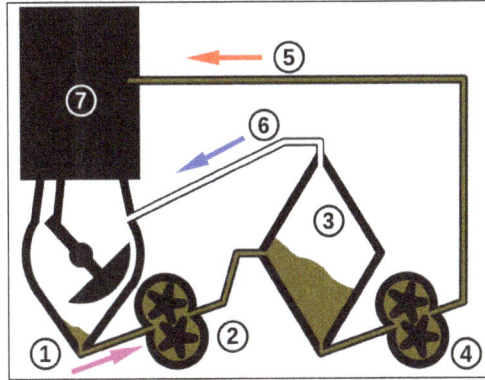

In the above schematic diagram of a basic dry-sump engine lubrication system. The oil collects in sump (1), is withdrawn continuously by scavenge pump (2) and travels to the oil tank (3), where gases entrained in the oil separate and the oil cools. Gases (6) are returned to the engine sump. Pressure pump (4) forces the de-gassed and cooled oil (5) back to the engine's lubrication points (7).

Engines are both lubricated and cooled by oil that circulates throughout the engine, feeding various bearings and other moving parts and then draining, via gravity, into the sump at the base of the engine. In the wet-sump system of most production automobile engines, a pump collects this oil from the sump and directly circulates it back through the engine. In a dry-sump, the oil still falls to the base of the engine, however instead of collecting in a reservoir-style oil sump, it falls into a much shallower sump, where one or more scavenge pumps draw it away and transfer it to a (usually external) reservoir, where it is both cooled and de-aerated before being recirculated through the engine by a pressure pump. Dry-sump designs frequently mount the pressure pump and scavenge pumps on a common crankshaft, so that one pulley at the front of the system can run as many pumps as the engine design requires. It is common practice to have one scavenge pump per crankcase section, however in the case of inverted engines (aircraft engine) it is necessary to employ separate scavenge pumps for each cylinder bank. Therefore, an inverted V engine would have a minimum of two scavenge pumps and a pressure pump in the pump *stack*.

The main purpose of the dry-sump system is to contain all the stored oil in a separate tank, or reservoir. This reservoir is usually tall and round or narrow and specially designed with internal baffles, and an oil outlet (supply) at the very bottom for uninhibited oil supply. The dry-sump oil pump is a minimum of 2 stages, with as many as 5 or 6. One stage is for pressure and is supplied the oil from the bottom of the reservoir, and along with an adjustable pressure regulator, supplies the oil under pressure through the filter and into the engine. The remaining stages "scavenge" the oil out of the dry-sump pan and return the oil (and gasses) to the top of the tank or reservoir. If an oil cooler is used usually it is mounted inline between the scavenge outlets and the tank. The dry-sump pump is usually driven by a Gilmer or HTD timing belt and pulleys, off the front of the crankshaft, at approximately one half crank speed. The dry-sump pump is designed with multiple stages to ensure that all the oil is scavenged from the pan and also to allow removal of excess air from the crankcase.

Advantages

A dry-sump system offers many advantages over a wet-sump. The primary advantages include:

- Improved engine reliability due to consistent oil pressure. This is the reason why dry-sumps were invented.

- Increased oil capacity by using a large external reservoir, that would be impractical in a wet-sump system.

- Prevention of the engine experiencing oil-starvation during high g-loads, which is particularly useful in racing cars, high performance sports cars, and aerobatic aircraft. Dry-sump designs are not susceptible to the oil movement problems from high cornering forces that wet-sump systems can suffer where the force of the vehicle cornering can cause the oil to pool on one side of the oil pan, possibly uncovering the oil pump pickup tube and causing cavitation and loss of oil pressure.

- Improvements to vehicle handling and stability. The vehicle's center of gravity can be lowered by mounting the engine lower in the chassis due to a shallow sump profile. A vehicle's overall weight distribution can be modified by locating the external oil reservoir away from the engine.

- Improved oil temperature control. This is due to increased oil volume providing resistance to heat saturation, the positioning of the oil reservoir away from the hot engine, and the ability to include cooling capabilities between the scavenger pumps and oil reservoir and also within the reservoir itself.

- The ability to release gasses trapped in the oil from ring blow-by and the action of the crankshaft and other moving parts in the oil, then return these gases via a line from the top of the oil reservoir to the combustion chamber.

- Improved pump efficiency to maintain oil supply to the engine. Since scavenge pumps are typically mounted at the lowest point on the engine, the oil flows into the pump intake by gravity rather than having to be lifted up into the intake of the pump as in a wet-sump. Furthermore, scavenge pumps can be of a design that is more tolerant of entrapped gasses than the typical pressure pump, which can lose suction if too much air mixes into the oil. Since the pressure pump is typically lower than the external oil tank, it always has a positive pressure on its suction regardless of cornering forces.

- Increased engine horsepower due to reduced viscous and air friction. In a wet-sump engine the crank shaft and other moving parts splash through the oil at thousands of RPM causing a "hurricane that whips the oil in a wet-sump engine into an aerated froth like a milkshake in a blender". Additionally, in a wet-sump, each revolution generates minute amounts of parasitic power loss caused by viscous drag and air drag (or 'windage') as the parts move rapidly through the oil and air in the lower engine. At high RPM these small sources of drag compound dramatically, resulting in significant power loss. In a dry-sump, the scavenger pump removes the oil and therefore the source of viscous friction, but also creates an air vacuum that significantly reduces air-friction, thus freeing the moving parts of much viscous and air friction and allowing engine power output to increase.

- Having the pumps external to the engine makes them easier to maintain or replace.

Disadvantages

Dry-sump engines have several disadvantages compared to wet-sump engines, including;

- Dry-sump systems add cost, complexity, and weight.

- The extra pumps and lines in dry-sump engines require additional oil and maintenance.

- The performance-enhancing features of dry-sump lubrication can hurt a car's day-to-day driveability. A good example is the classic Mercedes-Benz 300SL, a car that was designed for racing but sold to the general public and used on-road. The car had high oil capacity and a dry-sump system to cope with continuous high-speed running while racing. Owners found in general use, however, that the oil never achieved the correct operating temperature because the system was so efficient at cooling the oil. A makeshift solution was devised to deliberately block the oil cooler airflow to boost the oil temperature.

- The large external reservoir and pumps can be tricky to position around the engine and within the engine bay due to their size.

- As wrist pins and pistons rely for lubrication and cooling respectively, on the oil being splashed around in the crankcase, these parts might have inadequate oiling if too much oil is pulled away by the pump. Installing piston oilers can circumvent this issue, but do so with additional cost and complexity for the engine.

- Inadequate upper valvetrain lubrication can also become an issue if too much oil vapor is being pulled out from the area, especially with multi-staged pumps.

Common Engine Applications

Dry-sumps are common on larger diesel engines such as those used for ship propulsion, largely due to increased reliability and serviceability. They are also commonly used in racing cars and aerobatic aircraft, due to problems with g-forces, reliable oil supply, power output and vehicle handling. The Chevrolet Corvette Z06 has a dry sump engine which requires initial oil change after 500 miles.

Motorcycle Engines

The dry-sump lubrication is particularly applicable to motorcycles, which tend to be operated more vigorously than other road vehicles. Although motorcycles such as the Honda CB750 feature a dry-sump engine, modern motorcycles tend to use a wet-sump design. This is understandable with across-the-frame inline four-cylinder engines, since these wide engines must be mounted fairly high in the frame (for ground clearance), so the space below may as well be used for a wet-sump. However, narrower engines can be mounted lower and ideally should use dry-sump lubrication.

Several motorcycle models that use dry-sumps include;

- The classic British parallel twin motorcycles, such as BSA, Triumph and Norton, all used dry-sump lubrication. Traditionally, the oil tank was a remote item, but some late-model BSAs, and the Meriden Triumphs, used "oil-in-the-frame" designs.

- The Yamaha TRX850 270-degree parallel twin motorcycle has a dry-sump engine. Its oil reservoir is not remote, but integral to the engine, sitting atop the gearbox. This design eliminates external oil lines, allowing simpler engine removal and providing faster oil warm up.

- The Yamaha XT660Z (and R/X models) use a dry-sump design where the bike's frame tubing is used as the oil reservoir and cooling system.

- The Yamaha SR400/500 uses a dry-sump design where the bike's frame tubing doubles as the oil reservoir and cooling system.

- Harley-Davidson has used dry-sump type lubricating oil systems in their engines since the 1930s.

- The Rotax engined Aprilia RSV Mille, and the Aprilia RST1000 Futura both incorporate a dry-sump, along with sister bikes, the SL1000 Falco and ETV1000 Caponord.

- The Honda NX650, XR500R, XR600R, XR650R and XR650L four-stroke dirt bikes utilize a dry-sump with the oil in the frame tubing.

- The Suzuki DR-Z400 has a 2L dry-sump with oil in the frame tubing.

- Chennai built Royal Enfield prior to 2007. Royal Enfield dry sump designs were completely phased out by 2012.

Wet Sump

A wet sump is a lubricating oil management design for piston engines which uses the crankcase as a built-in reservoir for oil, as opposed to an external or secondary reservoir used in a dry sump design.

Piston engines are lubricated by oil which is pumped into various bearings, and thereafter allowed to drain to the base of the engine under gravity. In most production automobiles and motorcycles, which use a wet sump system, the oil is collected in a 3 to 10 litres (0.66 to 2.20 imp gal; 0.79 to 2.64 US gal) capacity pan at the base of the engine, known as the sump or oil pan, where it is pumped back up to the bearings by the oil pump, internal to the engine.

A wet sump offers the advantage of a simple design, using a single pump and no external reservoir. Since the sump is internal, there is no need for hoses or tubes connecting the engine to an external sump which may leak. An internal oil pump is generally more difficult to replace, but that is dependent on the engine design.

A wet sump design can be problematic in a racing car, as the large g force pulled by drivers going around corners causes the oil in the pan to slosh, gravitating away from the oil pick-up, briefly starving the system of oil and damaging the engine. However, on a motorcycle this difficulty does not arise, as a bike leans into corners and the oil is not displaced sideways. Nevertheless, racing motorcycles usually benefit from dry sump lubrication, as this allows the engine to be mounted lower in the frame; and a remote oil tank can permit better lubricant cooling.

Early stationary engines employed a small scoop on the extremity of the crankshaft or connecting rod to assist with the lubrication of the cylinder walls by means of a splashing action. Modern small engines, such as those used in lawnmowers, use a "slinger" (basically a paddle wheel) to perform the same function.

Two-stroke Engines

Small two-stroke engines, as for motorcycles and lawnmowers, use crankcase compression: the fuel mixture passes through the sump space in the crankcase. This precludes the use of both wet sump and dry sump systems, as excess oil here would contaminate the mixture, leading to excess oil being burned in the engine and so excessive hydrocarbon emissions. These engines are instead lubricated by petroil mixtures, where a carefully measured proportion of oil is added to the fuel tank (between 1:16 and 1:50 ratios). This oil is consumed immediately and entirely, so there is no need for a sump to collect and re-use it.

Four-stroke engines, as for almost all cars, and large two-stroke engines used in locomotives and ships can both use either wet or dry sumps. Large two-stroke engines do not use crankcase compression; instead they use a separate blower or supercharger, either a mechanical blower such as a Roots blower or else a turbocharger.

Types of Wet Sump

- The Splash System

- The Splash and Pressure System

- The full Pressure Feed System

Ring Oiler

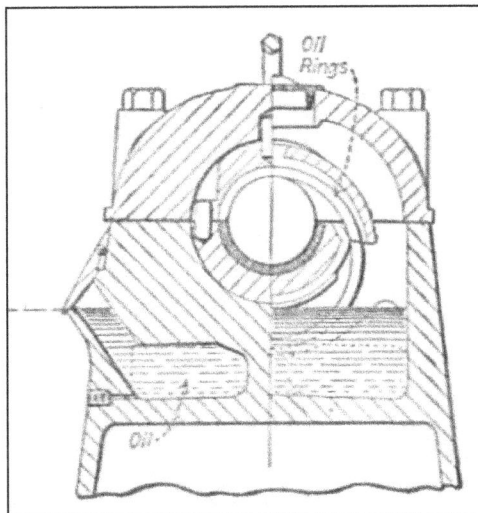

Section through a bearing, showing the oil sump beneath
and the ring oiler in place around the shaft.

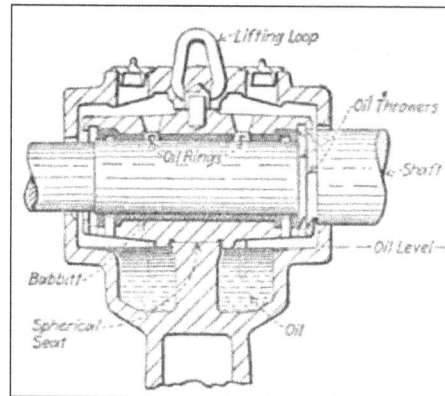

Section though a long Babbitt metal sleeve bearing, with two ring oilers
fitted through grooves in the upper part of the bearing.

A Ring Oiler or Oil Ring is a form of oil-lubrication System for Bearings. Ring oilers were used for medium-speed applications with moderate loads, during the first half of the 20th century. These represented the later years of the stationary steam engine, and the beginnings of the high-speed steam engine, the internal combustion oil engine and electrical generating equipment. Before this time plain bearings were lubricated by drip-feed oil cups or manually by an engine tender with an oil can. As speeds or bearing loads later increased, forced pressure lubrication became more prevalent and the ring oiler fell from use.

A ring oiler is a simple device, consisting of a large metal ring placed around a horizontal shaft, adjacent to a bearing. An oil sump is underneath this shaft and the ring is large enough to dip into the oil. As the shaft rotates, the ring is carried round with it. The rotating ring in turn picks up some oil and deposits it onto the shaft, from where it flows sideways and lubricates the bearings. The oil ring is effectively a simple lubrication pump, with only one moving part and no complex or high-precision components. The device is crude, but automatic, effective and reliable. Unlike a drip oiler, there is also no need to close off the oiler or remove oil wicks when the machine is stopped.

Ring oilers were used for speeds up to around 1,000 rpm. Above this, the oil tended to be thrown centrifugally from the ring, rather than carried by it (although it is still currently applied on steam turbines with speeds around 3200 rpm). The bearing must also remain horizontal and stable, so although suitable for crankshaft main bearings, they could not be used on connecting rod big end bearings. They were not used on vehicles for similar reasons, although the engines concerned at this time were anyway too large and heavy for practical mobile use. Automatic ring oilers were particularly useful for large engines with multiple horizontally opposed cylinders, where it was otherwise difficult to access the central main bearings. Ring oilers were most suited where bearing side-loads were relatively light, but the bearing capacity required more lubrication than could be supplied by a drip feed oiler. For this reason they were widely used on larger electric motors and generators.

Total-Loss Oiling System

A total-loss oiling system is an engine lubrication system whereby oil is introduced into the engine, and then either burned or ejected overboard. Now rare in four-stroke engines, total loss oiling is still used in many two-stroke engines.

Sight-glass lubricator.

Steam Engines

Steam engines used many separate oil boxes, dotted around the engine. Each one was filled before starting and often refilled during running. Where access was difficult, usually because the oil box was on a moving component, the oil box had to be large enough to contain enough oil for a long working shift. To control the flow rate of oil from the reservoir to the bearing, the oil would flow through an oil wick by capillary action, rather than downwards under gravity.

On steamships that ran their engines for days at a time, some crew members would be "oilers" whose primary duty was to continuously monitor and maintain oil boxes.

Displacement Lubricator for Adding Oil to a Steam Supply

On steam locomotives, access would be impossible during running, so in some cases centralised mechanical lubricators were used. These devices comprised a large oil tank with a multiple-outlet pump which fed the engine's bearings through a pipe system. Lubrication of the engine's internal valves was done by adding oil to the steam supply, using a displacement lubricator.

Oil Recirculation

The first recirculating systems used a collection sump, but no pumped circulation, merely 'splash' lubrication where the connecting rod dipped into the oil surface and splashed it around. These first appeared on high-speed steam engines. Later, splash lubrication engines added a 'dipper', a metal rod whose only function was to dip into the oil and spread it around.

As engines became faster and more powerful, the amount of oil required became so great that a total loss system would have been impractical, both technically and for cost.

Splash lubrication was also used on the first internal combustion engines. It persisted for some time, even in the first high-performance cars. One of Ettore Bugatti's first technical innovations was a minor improvement to the splash lubrication of crankshafts, helping to establish his reputation as an innovative engineer.

A more sophisticated form of splash lubrication, long-used for rotating motor shafts rather than reciprocating engines, was the ring oiler.

Pumped Oil

Later systems collect oil in a sump, from where it can be collected and pumped around the engine again, usually after rudimentary filtering. This system has long been the norm for larger internal combustion engines.

A pumped oil system can use higher oil pressures and so makes the use of hydrostatic bearings easier. These gave a greater load capacity and soon became essential for small, lightweight engines such as in cars. It was this bearing design that saw the end of splash lubrication and total loss oiling. It disappeared from nearly all cars in the 1920s, although total loss continued in small low power stationary engines into the 1950s. Chevrolet used splash lubrication for their rod bearings until 1953, where it was phased out for the 235 'Six,' and then in 1954 when the 216 was eliminated from their line, and both the solid lifter and hydraulic lifter versions of the 235 had full-pressure lubrication.

Two Stroke Engines and Petroil Mixtures

Two-stroke engines, and most model engines, have a total-loss lubrication system. Lubricating oil is mixed with the fuel, either manually beforehand (the petroil method), or automatically via an oil pump. Prior to being burned in the combustion chamber, this air/fuel/oil mixture passes through the engine's crankcase, lubricating the moving parts as it does so. In order to reduce exhaust smoke, the Kawasaki H2 750 cc (46 cu in) 2-stroke triple motorcycle had a scavenge pump with a spring-loaded ball-valve under each crankcase to return surplus oil to the tank for reuse.

Rotating-Crankcase Radial Engines

Normally known by the term "rotary engine", the usually air-cooled radial configuration, rotating-crankcase Otto cycle engines used by many "pioneer era" aircraft, and World War I combat aircraft used this particular type of engine, inherently designed to have a total-loss lubrication system, with the motor oil held in a separate tank from the fuel in the vehicle, and *not* pre-mixed with it as with two-cycle engines, but mixed within the engine instead while running.

Wankel Engines

Wankel engines are internal combustion engines using an eccentric rotary design to convert pressure into rotating motion. These engines exhibit some features of both four stroke and two stroke engines. Lubrication is total loss, but there may be some variations. For instance, the MidWest AE series of wankel aero-engines were not only both water-cooled and air-cooled, but also the engine had a lubrication system is a semi-total-loss system. Silkolene 2-stroke oil was directly injected into the inlet tracts and onto the main roller bearings. The oil that entered the combustion chamber lubricated the rotor tips and was then total-loss, but the oil that fed the bearings became a mist within the rotor-cooling air, and around 30% of that oil was recovered and returned to the remote oil tank.

Oil Mist Lubrication System

An oil mist system is a centralized lubrication system that generates, conveys and automatically delivers lubricant to machinery bearings. It is a system that has few moving parts making it very reliable. The lean mixture of oil and air produced by the generator is known as oil mist. The oil particles form a stable suspension that can be conveyed considerable distance (180 meters) through piping and tubing to the point requiring lubrication. Oil mist is a proven technology and it provides many advantages over conventional lubrication techniques such as oil splash and grease.

Applications and Use

Oil mist is used to lubricate rolling element bearings of all types. The most common applications in refineries and petrochemical plants are the bearings in pumps and their electric motor drivers. In addition, oil mist is used to purge gearboxes and the bearing houses of small steam turbines using sleeve bearings. Oil mist systems have been used in the hydrocarbon processing industry since the late 1960's with widespread use in many areas of the world commencing in the 1980's.

Financial Benefits

Oil mist systems are justified for economic reasons. It has been documented in technical papers released by users, by bearing manufacturers and in publications on university research that bearings lubricated with oil mist have longer life than bearings lubricated with oil sump or grease. Users claim from 50% to 90% reduction in lubrication related bearing failures when oil mist is used. A bearing failure can lead to an equipment overhaul costing $5,000 to $10,000, not including lost production. Reduction in seal failures has also been documented with the use of oil mist. The savings result in payback on the investment in less than two years and this combined with twenty-year life and low system operating cost means oil mist is a high-return, low-risk project.

Reasons for Superior Performance

- Bearings run cooler, typically on the order of 10°C. Lower bearing operating temperature means longer life.

- Oil mist lubrication is contaminant free. Studies have shown that it is common to have high contamination levels in oil sumps even though bearing housing seals are in place and recommended oil change intervals are followed. This contamination is not present with oil mist.

- External contaminants such as water vapor and particulate matter are excluded from the bearing housing because the mist system operates at positive pressure.

- The internal surfaces of the bearing housing are always coated with oil so there is no possibility for corrosion. This also applies to stand-by equipment meaning back-up machinery is ready for operation.

Oil Mist and Industry Specifications

The American Petroleum Institute (API) and the Process Industry Practices (PIP) group have taken a position that supports the use of oil mist. In their "Recommended Practices for Machinery Installation and Installation Design" specification, oil mist systems are prominently described. In the API specification for pump design (API-610) oil mist lubrication is included as a means for lubrication.

Pure Oil Mist Application for Pump API and Motor

- Pure Oil Mist provides the lubrication, continually supplies to bearing.

- Bearing run cooler typically 10 °C, lower friction, less heat generation.

- Wear Particles are Continuously Washed from the Bearing Surface.

Purge Oil Mist Application for Turbine

- Oil Mist purge over oil level in housing with slight positive pressure to keep oil free from contamination.

- Continuously coats all internal surfaces with Lubricant to anti corrosive agent.

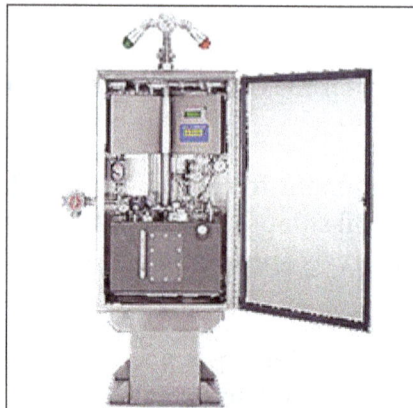

Oil Mist Equipment Designs and Certifications

The microprocessor controlled central consoles are easily tied into DCS systems. In addition, consoles that are third party approved for NEC and IEC hazardous area classifications are available. It is not necessary to carry out special designs for particular projects as the vendor equipment is already approved for use in hazardous classified areas.

Machinery Manufactures

As the use of oil mist has grown around the world, pump and electric motor manufacturers have incorporated bearing housing designs for oil mist lubrication. Pump and electric motor manufacturers that have a significant presence in the hydrocarbon processing industry supply equipment ready for oil mist lubrication. API-610 gives them clear standards for use of oil mist with pumps. The procedures for converting from grease lubricated motor bearings to oil mist are also readily available and easily adapted.

Oil mist lubrication is a proven technology that delivers both financial and environmental benefits. Equipment failures are reduced lowering operating and maintenance costs. Consumption and discharge of lube oil is also reduced with oil mist. Financial payback on the investment in oil mist systems is quick and low risk. Leading vendors of oil mist systems have the experience and knowledge for the successful application of oil mist to machinery in the hydrocarbon processing industry. The use of oil mist lubrication is becoming a preferred approach to bearing lubrication by many of the leading companies in this industry.

Automatic Lubrication

Automatic lubrication (also called autolube or auto-lube) refers to a lubrication system on a two-stroke engine, in which the oil is automatically mixed with fuel and manual oil-fuel pre-mixing is not necessary. The oil is contained in a reservoir that connects to a small oil pump in the engine, which needs to be periodically refilled.

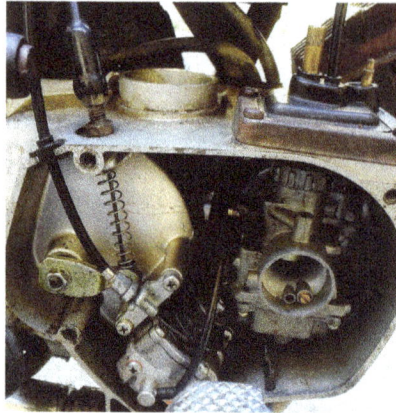

Oil injection pump on a Yamaha DX100- just behind the carburettor (visible on the left). It is the primary component of two-stroke Automatic Lubrication System. Amount of two-stroke oil injected by the pump depends on the throttle position. A cable from the throttle is connected to the oil pump indicating throttle's position. A tube ensures flow of oil from the reservoir to the oil pump.

This system is commonly used for motorcycles as it eliminates the need of pre-mixing fuel and two-stroke oil. Vespa Sprint is an example where pre-mixing of two-stroke oil is required. Automatic lubrication was introduced for motorcycles by Velocette in 1913.

An example of application of automatic lubrication system is Suzuki AX100 motorcycle. The motorcycle has a separate oil reservoir on its right side which supplies the cylinder with two-stroke oil proportional to engine speed.

Advantages

- Consistent lubrication and oil consumption is reduced greatly.

- More effective lubrication results because the oil enters the engine in larger size droplets.

- There is much less unwanted carbon deposited on the spark plugs, cylinder heads, pistons and exhaust system.

- There is much less exhaust smoke.

- Refueling is simplified.

Disdvantages

- The system is more complicated compared to manual pre-mixing, although it is easier for the end user.

- For any reason, if the two-stroke oil pump fails to operate properly, chance of damaging the engine is very high.

- The two-stroke oil tank in scooters and motorcycles is usually hidden from direct view of the rider and needs filling up occasionally. Without any indicator to indicate oil level, it is possible for a novice rider to forget to fill up the oil tank. This can end up starving the engine of oil and cause damage.

Each year millions of dollars are spent for new plant equipment designed to build things better and faster. However, machines continue to break down.

Bearing failure is a major cause of equipment downtime in today's industrial environment, most often resulting from improper lubrication. Improper lubrication scenarios include the contamination of the lubricant by dust, dirt and moisture, inadequate amounts of lubricant applied to the bearing, and/or overlubrication of the bearing.

Each bearing failure directly impacts the production cycle. While bearings can be expensive, replacement cost alone is often miniscule compared to lost production and the cost to repair the damage.

Why, in the age of technology, is this a problem? It is because many bearings are still lubricated manually. No matter how diligently a maintenance staff adheres to a lubrication schedule, it is a difficult task. Because employees are expected to manage multiple responsibilities in the lean environment of today's plants, it is common that proper lubrication is not considered a priority.

Benefits of Automatic Lubrication

Every bearing, regardless of size or location, needs to be lubricated properly. Improper lubrication will result in high, yet unnecessary costs for the operation. Some of the direct costs resulting from improper lubrication include replacement bearings, labor for replacements and repair, excess lubricant and labor for inefficient manual practice. Some of the indirect, but very real costs are downtime or lost production; product spoilage due to excess lubricant; environmental, safety or housekeeping issues; and excess energy consumption.

While grease guns and manual lubrication seem to get the job done for many maintenance operations, their benefits often cannot compare to those provided by an automatic lubrication system in terms of productivity, environmental issues and worker safety. An automatic lubrication system helps to prevent bearing failure by providing the right amount of the right (fresh, clean) lubricant at the right time to the right place.

The main difference between automatic and manual lubrication is that in the case of manually applied lubricants, technicians tend to lubricate on schedule (once a day, week, month, year, etc.) rather than when the bearing needs it. To compensate, the operator often will fill the bearing until he sees lubricant seeping out. The lubricant could be effectively "spent" by the time the operator gets back to it again. This sets up an overlubrication and underlubrication scenario. Conversely, automatic lubrication provides lubricant constantly at an appropriate amount that allows the bearing to operate at its optimum. When the bearing is properly lubricated in this manner, it also helps to seal the bearing from contaminants.

Maintaining proper lubrication on production equipment reduces the number of breakdowns due to bearing failure. In addition, there is less downtime due to the manual lubrication process, as well as substantially reduced man-hours required for the task.

Automatic lubrication is more precise and eliminates the cycle of overlubrication and underlubrication that contributes to bearing failure.

Manual Lube Safety. Even the most cautious maintenance personnel encounters
difficult conditions when performing manual lubrication, especially when machinery is in operation.

It also prevents excess lubricant from finding its way onto the finished product, the plant floor or other work surfaces. This results in fewer rejections, cleanup and disposal problems, as well as less waste of lubricant. And of course, all of this positively affects the company's bottom line.

Another benefit of an automatic lubrication system is worker safety. It becomes unnecessary for employees to engage in the potentially hazardous practice of manually applying lubricant while machinery is operating or in hazardous, difficult-to-reach locations.

Automatic Lubrication System

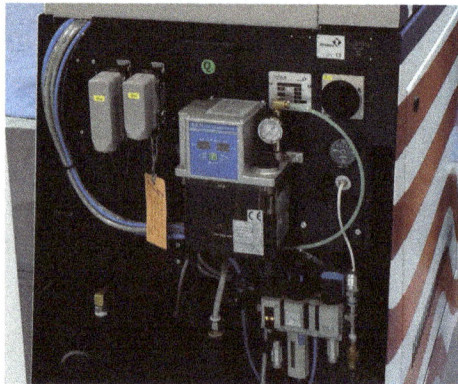

Automatic lubrication system installed on CNC machine.

An automatic lubrication system (ALS), often referred to as a centralized lubrication system, is a system that delivers controlled amounts of lubricant to multiple locations on a machine while the machine is operating. Even though these systems are usually fully automatic, a system that requires a manual pump or button activation is still identified as a centralized lubrication system. The system can be classified into two different categories that can share a lot of the same components.

Oil systems: Oil systems primary use is for stationary manufacturing equipment such as CNC milling.

Grease systems: Grease primary use is on mobile units such as trucks, mining or construction equipment.

Oil versus grease can vary even though their primary use is mostly stationary for oil and mobile for grease, some stationary manufacturing equipment will use grease systems.

Reason for an Automatic Lubrication System

Automatic lubrication system is designed to apply lubricant in small, measured amounts over short, frequent time intervals. Time and human resource constraints and sometimes the physical location on machine often makes it impractical to manually lubricate the points. As a result, production cycles, machine availability, and manpower availability dictate the intervals at which machinery is lubricated which is not optimal for the point requiring lubrication. Auto lube systems are installed on machinery to address this problem.

Benefits

Auto lube systems have many advantages over traditional methods of manual lubrication:

- All critical components are lubricated, regardless of location or ease of access.

- Lubrication occurs while the machinery is in operation causing the lubricant to be equally distributed within the bearing and increasing the machine's availability.

- Proper lubrication of critical components ensures safe operation of the machinery.

- Less wear on the components means extended component life, fewer breakdowns, reduced downtime, reduced replacement costs and reduced maintenance costs.

- Measured lubrication amounts means no wasted lubricant.

- Safety - No climbing around machinery or inaccessible areas (gases, exhaust, confined spaces, etc.)

- Lower energy consumption due to less friction.

- Increased overall productivity resulting from increase in machine availability and reduction in downtime due to breakdowns or general maintenance.

- In this system lubrication the engine parts are lubricated under pressure feed.

Components

A typical system consists of controller/timer, pump w/reservoir, supply line, metering valves, and feed lines. Regardless of the manufacturer or type of system, all automatic lubrication systems share these 5 main components:

- Controller/Timer – activates the system to distribute lubrication can be linked to a POS system.

- Pump with Reservoir – stores and provides the lubricant to the system.w

- Supply Line – line that connects the pump to the metering valves or injectors. The lubricant is pumped through this.

- Metering Valves/Injectors– component that measures/dispenses the lubricant to the application points.

- Feed lines - line that connects the metering valves or injectors to the application points. The lubricant is pumped through this.

Types of Automatic Lubrication System

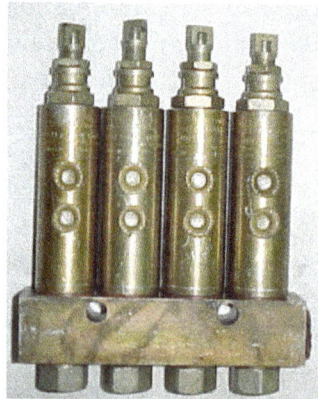

Single-line Injectors Installation. Lincoln Centro-Matic SI-1 on a large piece of mining machinery.

Once the decision to automate lubrication processes is made, there are several options from which to choose. Available systems include: single-line parallel, two-line parallel, single-line progressive, mist lubrication, minute-volume/low-pressure spray, recirculating oil, pump-to-point (box) lubricators, single-line resistance and single-point lubricators. A brief description of each follows.

Single-line Parallel

This system is easy to design, install, maintain, modify or expand. It typically operates at high fluid pressures and can be used with grease or oil. In this system, the pump pressurizes the main supply line, and a piston inside the primed injector displaces a premeasured amount of lubricant through the outlet to the bearing. The pump turns off and the supply line pressure is vented back to the reservoir. The spring-loaded piston returns to rest and the discharge chamber fills with a measured charge of lubricant for the next cycle.

Advantages

- Easy to design,

- Easy, cost-effective installation,

- Individually adjustable injectors,

- Proven, dependable design.

Disadvantages

- May not be suitable for combinations of heavy lubricants, very cold temperatures, very long supply line runs between pump and injectors.

Two-line Parallel

This system is ideal for long pumping distances and extreme temperatures. It is easy to adjust to meet specific bearing requirements. It provides high pressure, up to 5,000 psi, and is designed to work with many lubrication points over a wide area. This system's pump pressurizes metering devices through one side of a four-way, two-position reversing valve and the first supply line. The metering device control piston shifts and directs pressurized lubricant to the main piston, which displaces lubricant to the bearing. Lubricant on the other side of the control piston is vented back to the reservoir through the second supply line and the other side of the reversing valve. The reversing valve shifts and the pump pressurizes the second supply line, repeating the cycle in reverse.

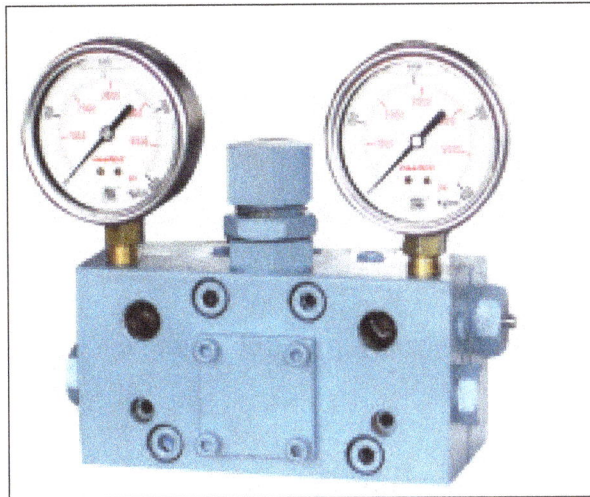

Two-line Reversing Valve. Directs the flow of lubricant from
one side of a two-line system to the other.

Advantages

- Easily handles very viscous (heavy) greases.

- Can accommodate long supply line runs between pump and metering devices.

Disadvantages

- May not be most cost effective for smaller systems.

- Requires two supply lines (another cost).

Single-line Progressive

This economical and flexible option is a system that can be used with low-pressure oil, grease or high-pressure oil. The latest designs include a preassembled pump, controller and mono-block piston-metering device. The pump of this type of system provides a measured single shot, pulsed or a continuous volume during the lubrication cycle. The first primed piston in the block shifts, displacing lubricant to the bearing and diverting flow to the next piston. The second piston shifts and diverts flow to the third. The sequence continues through the metering device until the timer or feedback switch stops the pump.

Progressive Divider Valve Installation. Installation can be completed using special, flexible lubrication line or high-pressure hydraulic pipe.

Advantages

- Accommodates a wide range of system control/monitoring options.

- Can identify blockage by monitoring a single point.

Disadvantages

- One blockage can disable the entire system.

- Large systems may require complex piping/tubing runs.

Mist Lubrication

Another simple system, mist lubrication facilitates low oil consumption and cool running bearings. Mist is generated with heat and/or air currents and is carried through pipe to the lubrication point with low-pressure air. Then it is sized to the appropriate droplet before it is dispensed to the bearing. Closed loop systems are environmentally friendly because they return the mist to the generator.

Advantages

- Cools and lubricates bearings.

- Low pressure keeps pipe material cost down.

- Positive pressure helps keep contaminants out of bearings.

Disadvantages

- Environmental/health concerns of "stray mist," especially with nonclosed loop systems.

- Oil only.

- Sensitivity to flow, viscosity, pressure variables.

- Extra pipe cost for closed loop systems.

Minute-Volume/Low-Pressure Spray

This system applies the precise amount of oil required by the lube point and has very low oil consumption, as much as 90 percent less than other methods. Environmentally friendly, this system is ideal for chain lubrication as it penetrates the wear points without overlubrication. When the timer signals the start of the lubrication event, the injector begins cycling, feeding oil at a controlled rate through small diameter tubing to the spray nozzle. Simultaneously, regulated low-pressure air is directed to the nozzle that mixes oil and air to produce a fine, controlled, nonmisting spray.

Advantages

- Precise lubricant volume and application control.

- No "stray mist" problems.

- Fast, economical installation.

- Very low lubricant consumption.

Disadvantages

- Oil only.

Multi-line, another pump-to-point technology, incorporates multiple metering devices that can be configured for a number of unique lubrication points and can be used for a wide variety of lubricants including heavy greases.

Recirculating Oil

This system is used to lubricate rolling element bearings and to maintain correct bearing temperature. It features a motor-driven pump that provides a continuous supply of oil through a filtration and piping system to flow meters. The flow meters control the amount of oil entering the bearing. The oil exits the bearing and returns to the reservoir through another piping system and return filter. Heat exchangers and/or heaters are used to maintain correct oil temperature. These systems are common on large, heavily loaded bearings in process industries.

Advantages

- Provides both lubrication and temperature control.

- Conditions, extends life of oil.

Disadvantages

- Most are major, capital installations.

- Some technologies require frequent manual adjustment at each flow meter.

Pump-to-point Lubricator

Handles multiple lubrication points independently and is ideal for remote locations. This system, which can overcome high back pressures, features an individual, adjustable lubricator pump for each point. Pump-to-point lubricators use an electric motor or machine power take-off to rotate a cam running through the drive box/reservoir. The cam actuates individual pump plungers through a rocker mechanism. A plunger then draws oil through a needle valve and sight glass and dispenses the measured volume though a high-pressure tube to the lubrication point. It is commonly used on large compressors and stationary gas engines.

Advantages

- Overcome extremely high backpressures.

- Simple, rugged design.

Disadvantages

- Limited number of lubrication points.

- Relatively high cost per point.

Single-line Resistance

A simple and cost-efficient system. Designed for closely clustered bearings, this system offers a variety of flow resistance metering devices and can utilize manual, electric or pneumatic pumps. The pump supplies a fixed volume of oil to the metering device through low-pressure tubing. The level of resistance in the metering device determines the proportion of oil flow to each lubrication point.

Advantages

- Simplicity.

- Low price.

Disadvantages

- Oil only.

- Reliance on resistance rather than piston metering devices can lead to nonpositive distribution of oil.

- System size limitations.

Single-point Lubricator

This is a simple and cost-efficient solution for individual, remote bearings. A completely self-contained unit, the single point lubricator is installed at each lubrication point. Gas pressure, spring or electromechanical power delivers the lubricant over time to the bearing. The reservoir or the entire unit is replaced when the lubricant is consumed, depending on the lubricator style.

Advantages

- Low purchase price.

- Easy to install.

Disadvantages

- Temperature will affect the volume output/service life of many single point lubricators.

- Replacement cost rapidly exceeds cost of fully automatic, central systems if the number of lube points increase.

Maintaining Automatic Lubrication Systems

Maintenance of automatic lubrication systems varies greatly from one system to the next. However, there are some common factors that can impact system performance and reliability.

First, the lubricant selected must be compatible with the intended application, the lubrication system components, system layout and ambient operating temperature range. When selecting a lubricant for a machine already equipped with automatic or centralized lubrication, make sure the lubricant grade is approved for use with the specific lubrication system.

When changing lubricant brands or types, be certain that the new lubricant is chemically compatible with the old. Although you may be adding the new lubricant to an empty and clean reservoir, residual old lubricant will be in the piping and metering devices. All systems require clean lubricant. Be sure to follow correct handling and storage procedures because a lubricant that is contaminated by dirt or moisture can damage lubrication system components, or worse, machine bearings.

A routine check of all fittings and piping is recommended. Leaks adversely affect lubrication system performance, as well as cause housekeeping, safety and/or environmental problems.

Keep system reservoirs filled with lubricant by topping them off on a routine basis. If the reservoir runs dry, most systems will pump air into the piping and metering devices. Air pockets are compressible and prevent systems from developing required pressure and volume of lubricant delivery, and the system will not function properly until the air is purged. Other maintenance issues vary, depending on the type of system in use. If the system is very basic, it may not have low-level indication or any other type of monitoring. These systems typically have transparent reservoirs or "sight glasses" and should be visually checked for lubricant level periodically.

On a single-line parallel system, the injectors have an indicator pin that comes out after each lubrication event so one can see that all injectors are cycling. A progressive system also may have indicator pins, allowing one to observe movement to verify system operation.

The larger single-line parallel systems use pressure switches or transducers in the supply line to indicate completion of a lubrication event. After reaching the prescribed pressure within an allocated period of time, a special controller resets and times out until the next lubrication cycle. If the system fails to reach the prescribed pressure, the controller activates an alarm. A limit switch attached to a cycle indicator pin can perform the same function on progressive systems.

Pressure-activated performance indicators also are a monitoring option on progressive systems. These devices, which mount parallel to the outlet port on the distribution block, are activated by high pressure, an indication of a blockage. When activated, the performance indicator either vents lubricant or extends a pin.

Many systems may include a reservoir low-level indication. With this feature, the system controller or independent indicator will signal the need to fill the reservoir. Many systems allow an audible alarm or signal light to make the alert more noticeable. Often, this same capability allows the user to route the alarm signal to the machine's PLC or to a central control panel in the plant.

In systems with sophisticated controls, remote sensors can be added to verify that each system cycle results in lubricant reaching the intended bearing. If the sensor does not detect lubricant flow, a fault signal is generated. This capability requires dedicated hardware and software to interpret the sensor signals.

References

- "Ford does not recommend API CK-4 or FA-4 oils in its diesel engines". Archived from the original on 9 December 2017. Retrieved 23 July 2017

- Engine-lubrication, automotive-lubrication, knowledge-center: masterlineworld.com, Retrieved 10 August, 2019

- Miller, S.J., N. Shan, and G.P. Huffman (2005). "Conversion of waste plastic to lubricating base oil". Energy & Fuels. 19 (4): 1580–6. Doi:10.1021/ef049696y

- The-working-of-an-engine-lubrication-system, machine-design: brighthubengineering.com, Retrieved 11 January, 2019

- Nunney, Malcom J. (2007). Light and Heavy Vehicle Technology (4th ed.). Elsevier Butterworth-Heinemann. P. 7. ISBN 978-0-7506-8037-0

- Oil-mist-lubrication-system, producteng: longwinthai.com, Retrieved 12 February, 2019

- Reher, David (2013-06-25). "Tech Talk #84 – Dry Sumps Save Lives". Reher Morrison Racing Engines. Retrieved 2016-12-24

- Automated-lubrication, Read: machinerylubrication.com, Retrieved 13 March, 2019

5

Technologies for Automotive Engines

Automotive engines make use of wide range of technologies such as air-cooling system, block heaters, radiators, antifreeze, turbochargers, boost controller, tuned exhaust, dual mass fly-wheel, etc. These technologies associated with automotive engines have been explained in this chapter.

Air Cooling System

In the air cooling system, the heat is dissipated directly to the air after being conducted through the cylinder walls. Air cooling systems have fins and flanges on the outer surfaces of the cylinders. The heads serve to increase the area exposed to the cooling air, and so raise the rate of cooling.

The basic principle involved in this method is to have a current of air flowing continuously over the heated surface of the engine from where the heat is to be removed. The amount of heat dissipated based on the following factors.

1. The surface area of metal into contact with air.

2. The rate of air flow.

3. A temperature difference between the heated surface and the air.

4. The conductivity of the metal.

For complete use of air-cooling, the surface area of the metal which comes in contact with air is improved by providing fins over the cylinder barrels. More the surface area in a contact with air, more the heat is dissipated. Higher the rate of air flow, higher the heat is dissipated.

Similarly higher the temperature difference between the heated surface and the air, higher will be the heat dissipation. A metal having conductivity dissipates more amount of heat.

Components of Air-Cooled Engines

The components of most air-cooling systems are very simple.

The cooling fan is placed in a semicircular ducting. The ducting covers the cylinder head. Its interior is fitted with baffles which direct the flow of air over the engine cooling fins and through an oil cooler.

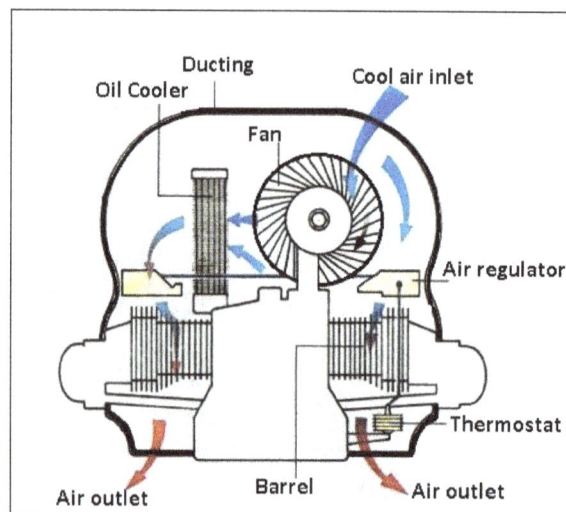

Below the cylinders, the air is delivered over a thermostat, which operates a valve via a lever. The valve controls the amount of air reaching the fan, thus maintaining the correct engine temperature. After passing over the engine and thermostat, the air is forced out of the rear of the car or passed through a heat transfer system that supplies hot water to the car's heater.

One problem connected with the use of air-cooled engines is the requirement of enough heating and demisting system for the car.

Water-cooled engines always have a constant supply of hot water and it is easy enough to convert this into hot air. Air-cooled engines usually have an independent heater or harness the heat of the exhaust system. Some older models have heating systems that combine both of these methods. An electrically-operated heater which burns petrol supplied hot air to the car interior by way of a blower fan. The same fan fed the hot air from the heat exchangers, which were finned alloy castings on the exhaust system. Hot air was fed into a mixer chamber, where it was blended with fresh air to give a controlled amount of heat.

Advantage of Air Cooling System Engine

- Lighter in weight due to there is no radiator, cooling jackets and coolant.

- No topping up the cooling system.

- No leaks to guard against.

- Anti-freeze not required.

- Engine warms-up faster than with water-cooled design.

- This system can be work in cold climates where water may freeze.

- Can be used in areas where there is a scarcity of cooling water.

Disadvantages of Air Cooling System Engine

- Less efficient cooling system, because of the coefficient of heat transfer for air is less than that for water.

- It is not easy to maintain even cool around the cylinder, the cylinder deformation can occur.

- More noisy operation.

- Limited use in motorcycles, and scooters where the cylinders are exposed to the air stream.

Cooling Fins

The surface area over the cylinder is get bigger by means of fins. These fins are either cast as an essential part of the cylinder or different finned barrels are placed over the cylinder barrel. Sometimes particularly in aero engines, the fins are machined from the forged cylinder blanks.

As a rule, the fins are usually made of about the cylinder wall thickness at their roots, tapering down to about one-half the root thickness. The length of the fins varies from one-quarter to one-third of the cylinder diameter. The distance between the two fin centres is about one-quarter to one-third of their length. The total length of the finned cylinder barrel is from 1 to 1½ times the cylinder bore.

Another rule based upon experimental considerations is to allow 1400 to 2400 cm² of cooling fins area per horsepower. This gives about the correct cylinder temperature at 50 to 70 km/hr air-speed.

Fan Cooling

Fan cooling is used in larger air-cooled engines, particularly on cars. A fan, having two or four blades, is driven either at engine speed or twice the engines speed, and the air-flow is directed in the cylinder heads. The cooling depends chiefly upon the engine speed and not upon the forward speed of the car. The fan usually absorbs about H.P. for every 15 to 20 H.P. output.

In the case of small single cylinder engines, an excellent arrangement of fan cooling is that a fan of about flywheel diameter. The fan is mounted on the main shaft and enclosed in a metal casing. So arranged that the air is drawn in at the centre and expelled peripherally through a belt mounted duct directing it on to the exhaust side of the cylinder.

In small air-cooled engines, the blower type fan works quite well, if suitable guides and ducts are provided for the air streams. The system is also used for the larger engine. The cooling system on the suction side of the fan, a more satisfactory cooling effect is obtained. Sometimes, the flywheel itself is tightened to function as a cooling fan. And the air is discharged backwards through it, after having been drawn past the cylinder barrels.

In fiat and Corvair engines, the quality of cooling air is regulated thermostatically. When the temperature of the air discharged from the cylinder increase above the normal value, the thermostat actuates a larger valve, or disc in the air outlet duct to allow a greater quality of air to flow.

Example of Air Cooling System in Engines

At present, air cooling is used on engines ex. like scooters, motorcycles, aeroplanes, combat tanks, small stationary installations. And in many models of an American rear-engine car. In Germany, air cooling is used in some petrol and C.I. engines including 2, 4 and 8 cylinder models.

A good example of the modern air-cooled type is the Krupp four-cylinder opposed compression ignition engine. This has a cooling fan fitted at the front end and is driven by the engine. It forced the cooling air through a casing around the front end of the crankcase and hence to the horizontal cylinder barrels which are ribbed and enclosed in a rectangular casing.

Another more recent example is the Krupp eight cylinder V-type petrol engine, Which has a very similar cooling arrangement.

The Volkswagon, Dutch D.A.F. Citroen-two-cylinder opposed. Chevrolet Corvair six-cylinder horizontally opposed, fiat 500D, two cylinders in line. And N.S.U. two cylinders are the example of modern air-cooled engines.

Block Heater

A block heater is used in cold climates to warm an engine prior to starting. They are mostly used for car engines, however they have also been used in aircraft engines. The most common design of block heater is an electrical heating element embedded in the engine block.

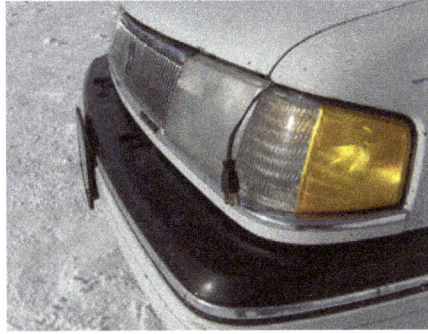

Electrical cord for powering a block heater.

Purpose

Pre-heating of an engine is primarily used to make it easier to start. It also results in the cabin heater producing heat sooner, both for the comfort of the driver and passengers and to de-fog the windscreen. An added benefit is that the condensation of fuel on cold metal surfaces inside the engine is reduced, thus an engine block heater reduces exhaust emissions during the start-up phase.

Block heaters or coolant heaters are also found on the diesel engines in standby generators, to reduce the time taken for the generator to reach full power output in an emergency.

Designs

Some cars are produced with block heaters from the factory, while others are fitted with block heaters as an aftermarket add-on. The most common type of block heater is an electric heating element in the engine block, which is connected through a power cord often routed through the vehicle's grille. Some block heaters are designed to replace one of the engine's core plugs and therefore heat the engine via the coolant.

Alternative methods of warming an engine include:

- Engine oil heater attached to the engine's oil pan with magnets.

- Engine oil heater inserted into the dipstick tube.

- In-line coolant heaters, which are installed into a radiator hose to warm the coolant (sometimes with a circulation pump to increase its effectiveness).

- Electric blankets that is laid over the top of the engine.

Electric timers are often used with engine warmers, since it is only necessary to run the warmer for a few hours before starting the engine. Some cars pump hot coolant from the cooling system into a 3-litre insulated thermos-style reservoir at shutdown, where it stays warm for several days.

Usage

Block heaters are frequently used in regions with cold winters such as the northern United States, Canada, Russia and Scandinavia. In some countries where block heaters are commonly used, car-parks are sometimes fitted with electrical outlets for powering the block heaters.

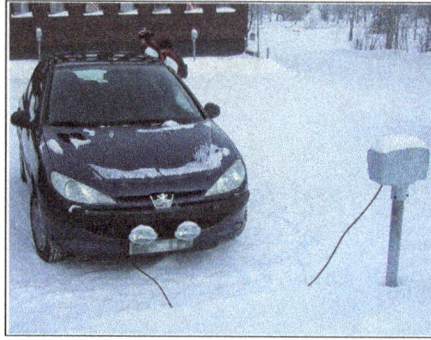

A parked car plugged in to an electrical outlet to power the block heater.

Testing in the 1970s of warm-up times for block heaters found little benefit in operating a block heater for than four hours prior to starting a vehicle. It was found that coolant temperature increased by almost 20 °C (36 °F) during the first four hours, regardless of the initial temperature. Four tests were run at ambient temperatures ranging from −29 to −11 °C (−20 to 12 °F); continued use of the heater for up to two hours more only further increased the temperature by up to 3 °C (5 °F). Engine oil temperature was found to increase over these periods by just 5 °C (9 °F).

Radiator

A typical engine coolant radiator used in an automobile.

Radiators are heat exchangers used for cooling internal combustion engines, mainly in automobiles but also in piston-engined aircraft, railway locomotives, motorcycles, stationary generating plant or any similar use of such an engine.

Internal combustion engines are often cooled by circulating a liquid called *engine coolant* through the engine block, where it is heated, then through a radiator where it loses heat to the atmosphere, and then returned to the engine. Engine coolant is usually water-based, but may also be oil. It is common to employ a water pump to force the engine coolant to circulate, and also for an axial fan to force air through the radiator.

Automobiles and Motorcycles

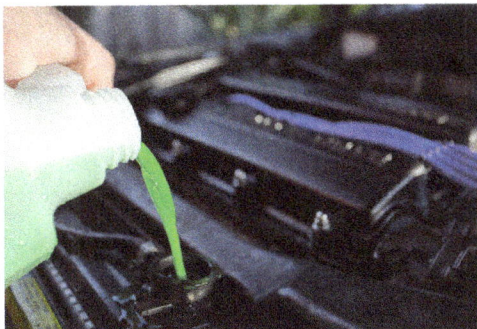

Coolant being poured into the radiator of an automobile.

In automobiles and motorcycles with a liquid-cooled internal combustion engine, a radiator is connected to channels running through the engine and cylinder head, through which a liquid (coolant) is pumped. This liquid may be water (in climates where water is unlikely to freeze), but is more commonly a mixture of water and antifreeze in proportions appropriate to the climate. Antifreeze itself is usually ethylene glycol or propylene glycol (with a small amount of corrosion inhibitor).

A typical automotive cooling system comprises:

- A series of galleries cast into the engine block and cylinder head, surrounding the combustion chambers with circulating liquid to carry away heat;

- A radiator, consisting of many small tubes equipped with a honeycomb of fins to dissipate heat rapidly, that receives and cools hot liquid from the engine;

- A water pump, usually of the centrifugal type, to circulate the coolant through the system;

- A thermostat to control temperature by varying the amount of coolant going to the radiator;

- A fan to draw cool air through the radiator.

The radiator transfers the heat from the fluid inside to the air outside, thereby cooling the fluid, which in turn cools the engine. Radiators are also often used to cool automatic transmission fluids, air conditioner refrigerant, intake air, and sometimes to cool motor oil or power steering fluid. Radiators are typically mounted in a position where they receive airflow from the forward movement of the vehicle, such as behind a front grill. Where engines are mid- or rear-mounted, it is common to mount the radiator behind a front grill to achieve sufficient airflow, even though this requires long coolant pipes. Alternatively, the radiator may draw air from the flow over the top of the vehicle or from a side-mounted grill. For long vehicles, such as buses, side airflow is most common for engine and transmission cooling and top airflow most common for air conditioner cooling.

Radiator Construction

Automobile radiators are constructed of a pair of metal or plastic header tanks, linked by a core with many narrow passageways, giving a high surface area relative to volume. This core is usually made of stacked layers of metal sheet, pressed to form channels and soldered or brazed together. For many years radiators were made from brass or copper cores soldered to brass headers. Modern radiators have aluminum cores, and often save money and weight by using plastic headers with gaskets. This construction is more prone to failure and less easily repaired than traditional materials.

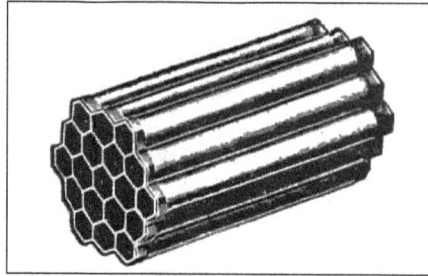

Honeycomb radiator tubes.

An earlier construction method was the honeycomb radiator. Round tubes were swaged into hexagons at their ends, then stacked together and soldered. As they only touched at their ends, this formed what became in effect a solid water tank with many air tubes through it.

Some vintage cars use radiator cores made from coiled tube, a less efficient but simpler construction.

Coolant Pump

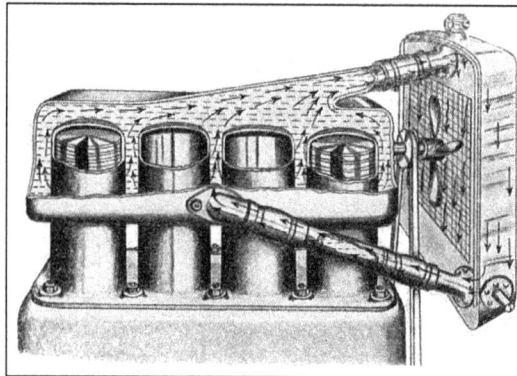

Thermosyphon cooling system of 1937, without circulating pump.

Radiators first used downward vertical flow, driven solely by a thermosyphon effect. Coolant is heated in the engine, becomes less dense, and so rises. As the radiator cools the fluid, the coolant becomes denser and falls. This effect is sufficient for low-power stationary engines, but inadequate for all but the earliest automobiles. All automobiles for many years have used centrifugal pumps to circulate the engine coolant because natural circulation has very low flow rates.

Heater

A system of valves or baffles, or both, is usually incorporated to simultaneously operate a small radiator inside the vehicle. This small radiator, and the associated blower fan, is called the heater core, and serves to warm the cabin interior. Like the radiator, the heater core acts by removing heat from the engine. For this reason, automotive technicians often advise operators to turn on the heater and set it to high if the engine is overheating, to assist the main radiator.

Temperature Control

Waterflow Control

Car engine thermostat.

The engine temperature on modern cars is primarily controlled by a wax-pellet type of thermostat, a valve which opens once the engine has reached its optimum operating temperature.

When the engine is cold, the thermostat is closed except for a small bypass flow so that the thermostat experiences changes to the coolant temperature as the engine warms up. Engine coolant is directed by the thermostat to the inlet of the circulating pump and is returned directly to the engine, bypassing the radiator. Directing water to circulate only through the engine allows the engine to reach optimum operating temperature as quickly as possible whilst avoiding localised "hot spots." Once the coolant reaches the thermostat's activation temperature, it opens, allowing water to flow through the radiator to prevent the temperature rising higher.

Once at optimum temperature, the thermostat controls the flow of engine coolant to the radiator so that the engine continues to operate at optimum temperature. Under peak load conditions, such as driving slowly up a steep hill whilst heavily laden on a hot day, the thermostat will be approaching fully open because the engine will be producing near to maximum power while the velocity of air flow across the radiator is low. (The velocity of air flow across the radiator has a major effect on its ability to dissipate heat.) Conversely, when cruising fast downhill on a motorway on a cold night on a light throttle, the thermostat will be nearly closed because the engine is producing little power, and the radiator is able to dissipate much more heat than the engine is producing. Allowing too much flow of coolant to the radiator would result in the engine being over cooled and operating at lower than optimum temperature, resulting in decreased fuel efficiency and increased exhaust emissions. Furthermore, engine durability, reliability, and longevity are sometimes compromised, if any components (such as the crankshaft bearings) are engineered to take thermal expansion into account to fit together with the correct clearances. Another side effect of over-cooling is reduced performance of the cabin heater, though in typical cases it still blows air at a considerably higher temperature than ambient.

The thermostat is therefore constantly moving throughout its range, responding to changes in vehicle operating load, speed and external temperature, to keep the engine at its optimum operating temperature.

On vintage cars you may find a bellows type thermostat, which has a corrugated bellows containing a volatile liquid such as alcohol or acetone. These types of thermostats do not work well at cooling system pressures above about 7 psi. Modern motor vehicles typically run at around 15 psi, which precludes the use of the bellows type thermostat. On direct air-cooled engines this is not a concern for the bellows thermostat that controls a flap valve in the air passages.

Airflow Control

Other factors influence the temperature of the engine, including radiator size and the type of radiator fan. The size of the radiator (and thus its cooling capacity) is chosen such that it can keep the engine at the design temperature under the most extreme conditions a vehicle is likely to encounter (such as climbing a mountain whilst fully loaded on a hot day).

Airflow speed through a radiator is a major influence on the heat it dissipates. Vehicle speed affects this, in rough proportion to the engine effort, thus giving crude self-regulatory feedback. Where an additional cooling fan is driven by the engine, this also tracks engine speed similarly.

Engine-driven fans are often regulated by a viscous-drive clutch from the drivebelt, which slips and reduces the fan speed at low temperatures. This improves fuel efficiency by not wasting power on driving the fan unnecessarily. On modern vehicles, further regulation of cooling rate is provided by either variable speed or cycling radiator fans. Electric fans are controlled by a thermostatic switch or the engine control unit. Electric fans also have the advantage of giving good airflow and cooling at low engine revs or when stationary, such as in slow-moving traffic.

Before the development of viscous-drive and electric fans, engines were fitted with simple fixed fans that drew air through the radiator at all times. Vehicles whose design required the installation of a large radiator to cope with heavy work at high temperatures, such as commercial vehicles and tractors would often run cool in cold weather under light loads, even with the presence of a thermostat, as the large radiator and fixed fan caused a rapid and significant drop in coolant temperature as soon as the thermostat opened. This problem can be solved by fitting a radiator blind (or radiator shroud) to the radiator that can be adjusted to partially or fully block the airflow through the radiator. At its simplest the blind is a roll of material such as canvas or rubber that is unfurled along the length of the radiator to cover the desired portion. Some older vehicles, like the World War I-era S.E.5 and SPAD S.XIII single-engined fighters, have a series of shutters that can be adjusted from the driver's or pilot's seat to provide a degree of control. Some modern cars have a series of shutters that are automatically opened and closed by the engine control unit to provide a balance of cooling and aerodynamics as needed.

Cooling fan of radiator for prime mover of a VIA Rail locomotive.

These AEC Regent III RT buses are fitted with radiator blinds,
seen here covering the lower half of the radiators.

Coolant Pressure

Because the thermal efficiency of internal combustion engines increases with internal temperature, the coolant is kept at higher-than-atmospheric pressure to increase its boiling point. A calibrated pressure-relief valve is usually incorporated in the radiator's fill cap. This pressure varies between models, but typically ranges from 4 to 30 psi (30 to 200 kPa).

As the coolant expands with increasing temperature, its pressure in the closed system must increase. Ultimately, the pressure relief valve opens, and excess fluid is dumped into an overflow container. Fluid overflow ceases when the thermostat modulates the rate of cooling to keep the temperature of the coolant at optimum. When the engine coolant cools and contracts (as conditions change or when the engine is switched off), the fluid is returned to the radiator through additional valving in the cap.

Engine Coolant

Before World War II, engine coolant was usually plain water. Antifreeze was used solely to control freezing, and this was often only done in cold weather.

Development in high-performance aircraft engines required improved coolants with higher boiling points, leading to the adoption of glycol or water-glycol mixtures. These led to the adoption of glycols for their antifreeze properties.

Since the development of aluminium or mixed-metal engines, corrosion inhibition has become even more important than antifreeze, and in all regions and seasons.

Boiling or Overheating

An overflow tank that runs dry may result in the coolant vaporizing, which can cause localized or general overheating of the engine. Severe damage can result, such as blown headgaskets, and warped or cracked cylinder heads or cylinder blocks. Sometimes there will be no warning, because the temperature sensor that provides data for the temperature gauge (either mechanical or electric) is exposed to air, not to the excessively hot coolant, providing a harmfully false reading.

Opening a hot radiator drops the system pressure, which may cause it to boil and eject dangerously hot liquid and steam. Therefore, radiator caps often contain a mechanism that attempts to relieve the internal pressure before the cap can be fully opened.

Supplementary Radiators

It is sometimes necessary for a car to be equipped with a second, or auxiliary, radiator to increase the cooling capacity, when the size of the original radiator cannot be increased. The second radiator is plumbed in series with the main radiator in the circuit. This was the case when the Audi 100 was first turbocharged creating the 200.

Some engines have an oil cooler, a separate small radiator to cool the engine oil. Cars with an automatic transmission often have extra connections to the radiator, allowing the transmission fluid to transfer its heat to the coolant in the radiator. These may be either oil-air radiators, as for a smaller version of the main radiator. More simply they may be oil-water coolers, where an oil pipe is inserted inside the water radiator. Though the water is hotter than the ambient air, its higher thermal conductivity offers comparable cooling (within limits) from a less complex and thus cheaper and more reliable oil cooler. Less commonly, power steering fluid, brake fluid, and other hydraulic fluids may be cooled by an auxiliary radiator on a vehicle.

Turbo charged or supercharged engines may have an intercooler, which is an air-to-air or air-to-water radiator used to cool the incoming air charge—not to cool the engine.

Aircraft

Aircraft with liquid-cooled piston engines (usually inline engines rather than radial) also require radiators. As airspeed is higher than for cars, these are efficiently cooled in flight, and so do not require large areas or cooling fans. Many high-performance aircraft however suffer extreme overheating problems when idling on the ground - a mere 7 minutes for a Spitfire. This is similar to Formula 1 cars of today, when stopped on the grid with engines running they require ducted air forced into their radiator pods to prevent overheating.

Surface Radiators

Reducing drag is a major goal in aircraft design, including the design of cooling systems. An early technique was to take advantage of an aircraft's abundant airflow to replace the honeycomb core (many surfaces, with a high ratio of surface to volume) by a surface mounted radiator. This uses a single surface blended into the fuselage or wing skin, with the coolant flowing through pipes at the back of this surface. Such designs were seen mostly on World War I aircraft.

As they are so dependent on airspeed, surface radiators are even more prone to overheating when ground-running. Racing aircraft such as the Supermarine S.6B, a racing seaplane with radiators built into the upper surfaces of its floats, have been described as "being flown on the temperature gauge" as the main limit on their performance.

Surface radiators have also been used by a few high-speed racing cars, such as Malcolm Campbell's Blue Bird of 1928.

Pressurized Cooling Systems

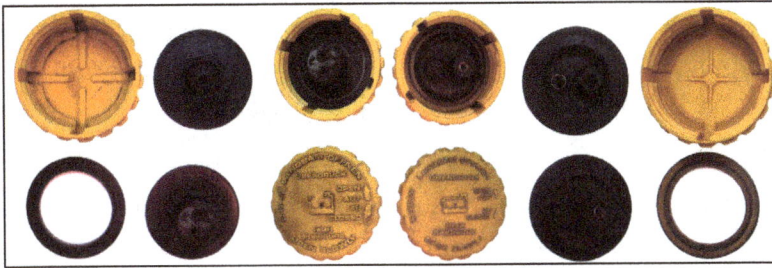

Radiator caps for pressurized automotive cooling systems. Of the two valves,
one prevents the creation of a vacuum, the other limits the pressure.

It is generally a limitation of most cooling systems that the cooling fluid not be allowed to boil, as the need to handle gas in the flow greatly complicates design. For a water cooled system, this means that the maximum amount of heat transfer is limited by the specific heat capacity of water and the difference in temperature between ambient and 100 °C. This provides more effective cooling in the winter, or at higher altitudes where the temperatures are low.

Another effect that is especially important in aircraft cooling is that the specific heat capacity changes with pressure, and this pressure changes more rapidly with altitude than the drop in temperature. Thus, generally, liquid cooling systems lose capacity as the aircraft climbs. This was a major limit on performance during the 1930s when the introduction of turbosuperchargers first allowed convenient travel at altitudes above 15,000 ft, and cooling design became a major area of research.

The most obvious, and common, solution to this problem was to run the entire cooling system under pressure. This maintained the specific heat capacity at a constant value, while the outside air temperature continued to drop. Such systems thus improved cooling capability as they climbed. For most uses, this solved the problem of cooling high-performance piston engines, and almost all liquid-cooled aircraft engines of the World War II period used this solution.

However, pressurized systems were also more complex, and far more susceptible to damage - as the cooling fluid was under pressure, even minor damage in the cooling system like a single rifle-calibre bullet hole, would cause the liquid to rapidly spray out of the hole. Failures of the cooling systems were, by far, the leading cause of engine failures.

Evaporative Cooling

Although it is more difficult to build an aircraft radiator that is able to handle steam, it is by no means impossible. The key requirement is to provide a system that condenses the steam back into liquid before passing it back into the pumps and completing the cooling loop. Such a system can take advantage of the specific heat of vaporization, which in the case of water is five times the specific heat capacity in the liquid form. Additional gains may be had by allowing the steam to become superheated. Such systems, known as evaporative coolers, were the topic of considerable research in the 1930s.

Consider two cooling systems that are otherwise similar, operating at an ambient air temperature of 20°C. An all-liquid design might operate between 30 °C and 90 °C, offering 60 °C of temperature difference to carry away heat. An evaporative cooling system might operate between 80°C and 110°C, which at first glance appears to be much less temperature difference, but this analysis overlooks the enormous amount of heat energy soaked up during the generation of steam, equivalent to 500°C. In effect, the evaporative version is operating between 80°C and 560°C, a 480°C effective temperature difference. Such a system can be effective even with much smaller amounts of water.

The downside to the evaporative cooling system is the *area* of the condensers required to cool the steam back below the boiling point. As steam is much less dense than water, a correspondingly larger surface area is needed to provide enough airflow to cool the steam back down. The Rolls-Royce Goshawk design of 1933 used conventional radiator-like condensers and this design proved to be a serious problem for drag. In Germany, the Günter brothers developed an alternative design combining evaporative cooling and surface radiators spread all over the aircraft wings, fuselage and even the rudder. Several aircraft were built using their design and set numerous performance records, notably the Heinkel He 119 and Heinkel He 100. However, these systems required numerous pumps to return the liquid from the spread-out radiators and proved to be extremely difficult to keep running properly, and were much more susceptible to battle damage. Efforts to develop this system had generally been abandoned by 1940. The need for evaporative cooling was soon to be negated by the widespread availability of ethylene glycol based coolants, which had a lower specific heat, but a much higher boiling point than water.

Radiator Thrust

An aircraft radiator contained in a duct heats the air passing through, causing the air to expand and gain velocity. This is called the Meredith effect, and high-performance piston aircraft with well-designed low-drag radiators (notably the P-51 Mustang) derive thrust from it. The thrust was significant enough to offset the drag of the duct the radiator was enclosed in and allowed the aircraft to achieve zero cooling drag. At one point, there were even plans to equip the Spitfire with an afterburner, by injecting fuel into the exhaust duct after the radiator and igniting it. Afterburning is achieved by injecting additional fuel into the engine downstream of the main combustion cycle.

Stationary Plant

Engines for stationary plant are normally cooled by radiators in the same way as automobile engines. However, in some cases, evaporative cooling is used via a cooling tower.

Cold Air Intake

A cold air intake (CAI) is usually an aftermarket assembly of parts used to bring relatively cool air into a car's internal-combustion engine.

Most vehicles manufactured from the mid-1970s until the mid-1990s have thermostatic air intake systems that regulate the temperature of the air entering the engine's intake tract, providing warm air when the engine is cold and cold air when the engine is warm to maximize performance, efficiency, and fuel economy. With the advent of advanced emission controls and more advanced fuel injection methods modern vehicles do not have a thermostatic air intake system and the factory installed air intake draws unregulated cold air. Aftermarket cold air intake systems are marketed with claims of increased engine efficiency and performance. The putative principle behind a cold air intake is that cooler air has a higher density, thus containing more oxygen per volume unit than warmer air.

Example of a Roush cold air intake system installed on a sixth generation Ford Mustang.

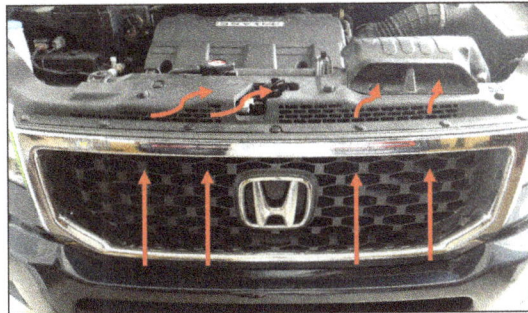

Illustration of how the first generation Honda Ridgeline's OEM cold air intake system gets fresh air forward of the radiator and into its airbox.

Design Features

Some strategies used in designing aftermarket cold air intakes are:

- Reworking parts of the intake that create turbulence to reduce air resistance.

- Providing a more direct route to the air intake by eliminating muffling devices.

- Shortening the length of the intake.

- Placing the intake duct so as to use the ram-air effect to give positive pressure at speed.

Construction

Intake systems come in many different styles and can be constructed from plastic, metal, rubber (silicone) or composite materials (fiberglass, carbon fiber or Kevlar). The most efficient intake systems utilize an airbox which is sized to complement the engine and will extend the powerband of the engine. The intake snorkel (opening for the intake air to enter the system) must be large enough to ensure sufficient air is available to the engine under all conditions from idle to full throttle.

The most basic cold air intake consists of a long metal or plastic tube leading to a conical air filter. Power may be lost at certain engine speeds and gained at others. Because of the reduced covering, intake noise is usually increased.

Some intakes use heat shields to isolate the air filter from the rest of the engine compartment, providing cooler air from the front or side of the engine bay. This can make a big difference to intake temperatures, especially when the car is moving slowly. Some systems, called "fender mount," move the filter into the fender wall instead. This system draws air up through the fender wall which provides even more isolation and still cooler air.

Antifreeze

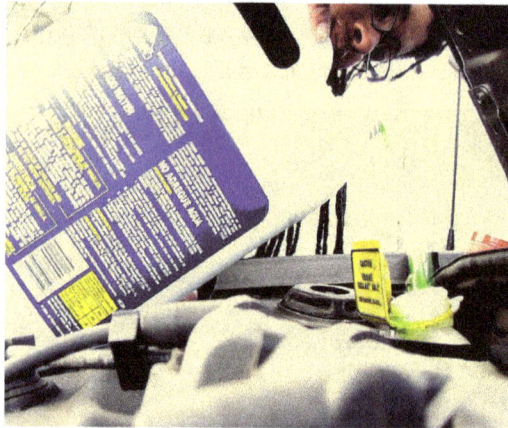

"Topping up" the antifreeze solution in a car's cooling system is a routine maintenance item for most modern cars.

An antifreeze is an additive which lowers the freezing point of a water-based liquid. An antifreeze mixture is used to achieve freezing-point depression for cold environments. Common antifreezes increase the boiling point of the liquid, allowing higher coolant temperature.

Because water has good properties as a coolant, water plus antifreeze is used in internal combustion engines and other heat transfer applications, such as HVAC chillers and solar water heaters. The purpose of antifreeze is to prevent a rigid enclosure from bursting due to expansion when water freezes. Commercially, both the *additive* (pure concentrate) and the *mixture* (diluted solution) are called antifreeze, depending on the context. Careful selection of an antifreeze can enable a wide temperature range in which the mixture remains in the liquid phase, which is critical to efficient heat transfer and the proper functioning of heat exchangers.

Water was the original coolant for internal combustion engines. It is cheap, nontoxic, and has a high heat capacity. It however has only a 100 °C liquid range, and it expands upon freezing. These problems are addressed by the development of alternative coolants with improved properties. Freezing and boiling points are colligative properties of a solution, which depend on the concentration of dissolved substances. Hence salts lower the melting points of aqueous solutions. Salts are indeed frequently used for de-icing, but salt solutions are not used for cooling systems because they induce corrosion of metals. Low molecular weight organic compounds tend to have melting points lower than water, which recommends them as antifreeze agents. Solutions of organic compounds, especially alcohols, in water are effective. Indeed alcohols - ethanol, methanol, ethylene glycol, etc - have been the basis of all antifreezes since they were commercialized in the 1920s.

Use and Occurrence

Automotive and Internal Combustion Engine use

Fluorescent green-dyed antifreeze is visible in the radiator header tank
when car radiator cap is removed.

Most automotive engines are "water"-cooled to remove waste heat, although the "water" is actually antifreeze/water mixture and not plain water. The term engine coolant is widely used in the automotive industry, which covers its primary function of convective heat transfer for internal combustion engines. When used in an automotive context, corrosion inhibitors are added to help protect vehicles' radiators, which often contain a range of electrochemically incompatible metals (aluminum, cast iron, copper, brass, solder, et cetera). Water pump seal lubricant is also added. When it comes to boats, there is no constant antifreeze. The type of antifreeze that is required for a certain motor boat depends on the engine location and design. Unlike antifreezes used for automotive application, assortment of boat antifreezes is more complex.

Antifreeze was developed to overcome the shortcomings of water as a heat transfer fluid.

On the other hand, if the engine coolant gets too hot, it might boil while inside the engine, causing voids (pockets of steam), leading to localized hot spots and the catastrophic failure of the engine. If plain water were to be used as an engine coolant, it would promote galvanic corrosion. Proper engine coolant and a pressurized coolant system obviate the shortcomings of water. With proper antifreeze, a wide temperature range can be tolerated by the engine coolant, such as -34 °F (-37 °C) to $+265$ °F (129 °C) for 50% (by volume) propylene glycol diluted with water and a 15 psi pressurized coolant system.

Early engine coolant antifreeze was methanol (methyl alcohol). Ethylene glycol was developed because its higher boiling point was more compatible with heating systems.

Other Industrial Uses

The most common water-based antifreeze solutions used in electronics cooling are mixtures of water and either ethylene glycol (EGW) or propylene glycol (PGW). The use of ethylene glycol has a longer history, especially in the automotive industry. However, EGW solutions formulated for the automotive industry often have silicate based rust inhibitors that can coat and/or clog heat exchanger surfaces. Ethylene glycol is listed as a toxic chemical requiring care in handling and disposal.

Ethylene glycol has desirable thermal properties, including a high boiling point, low freezing point, stability over a wide range of temperatures, and high specific heat and thermal conductivity. It also has a low viscosity and, therefore, reduced pumping requirements. Although EGW has more desirable physical properties than PGW, the latter coolant is used in applications where toxicity might be a concern. PGW is generally recognized as safe for use in food or food processing applications, and can also be used in enclosed spaces.

Similar mixtures are commonly used in HVAC and industrial heating or cooling systems as a high-capacity heat transfer medium. Many formulations have corrosion inhibitors, and it is expected that these chemicals will be replenished (manually or under automatic control) to keep expensive piping and equipment from corroding.

Biological Antifreezes

Antifreeze proteins refer to chemical compounds produced by certain animals, plants, and other organisms that prevent the formation of ice. In this way, these compounds allow their host organism to operate at temperatures well below the freezing point of water. AFPs bind to small ice crystals to inhibit growth and recrystallization of ice that would otherwise be fatal.

Primary Agents

Ethylene Glycol

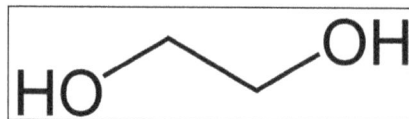

Ethylene Glycol.

Most antifreeze is made by mixing distilled water with additives and a base product - MEG (Mono ethylene glycol) or MPG (Mono propylene glycol). Ethylene glycol solutions became available in 1926 and were marketed as "permanent antifreeze" since the higher boiling points provided advantages for summertime use as well as during cold weather. They are used today for a variety of applications, including automobiles, but there are lower-toxicity alternatives made with propylene glycol available.

When ethylene glycol is used in a system, it may become oxidized to five organic acids (formic, oxalic, glycolic, glyoxalic and acetic acid). Inhibited ethylene glycol antifreeze mixes are available, with additives that buffer the pH and reserve alkalinity of the solution to prevent oxidation of ethylene glycol and formation of these acids. Nitrites, silicates, theodin, borates and azoles may also be used to prevent corrosive attack on metal.

Ethylene glycol poisoning is a problematic because this sweet-tasting chemical is toxic to humans and other animals.

Propylene Glycol

Propylene glycol is considerably less toxic than ethylene glycol and may be labeled as "non-toxic antifreeze". It is used as antifreeze where ethylene glycol would be inappropriate, such as in food-processing systems or in water pipes in homes where incidental ingestion may be possible. For example, the U.S. FDA allows propylene glycol to be added to a large number of processed foods, including ice cream, frozen custard, salad dressings, and baked goods, and it is commonly used as the main ingredient in the "e-liquid" used in electronic cigarettes.

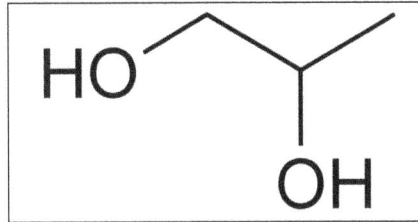

Propylene glycol.

Propylene glycol oxidizes to lactic acid. Besides cooling system corrosion, biological fouling also occurs. Once bacterial slime starts to grow, the corrosion rate of the system increases. Maintenance of systems using glycol solution includes regular monitoring of freeze protection, pH, specific gravity, inhibitor level, color, and biological contamination.

Propylene glycol should be replaced when it turns a reddish color. When an aqueous solution of propylene glycol in a cooling or heating system develops a reddish or black color, this indicates that iron in the system is corroding significantly. In the absence of inhibitors, propylene glycol can react with oxygen and metal ions, generating various compounds including organic acids (e.g., formic, oxalic, acetic). These acids accelerate the corrosion of metals in the system.

Other Antifreezes

Propylene glycol methyl ether is used as an antifreeze in diesel engines. It is more volatile than glycol.

Once used for automotive antifreeze, glycerol has the advantage of being non-toxic, withstands relatively high temperatures, and is noncorrosive. It is not however used widely. Glycerol was historically used as an antifreeze for automotive applications before being replaced by ethylene glycol. Glycerol is mandated for use as an antifreeze in many sprinkler systems.

Measuring the Freeze Point

Once antifreeze has been mixed with water and put into use, it periodically needs to be maintained. If engine coolant leaks, boils, or if the cooling system needs to be drained and refilled, the antifreeze's freeze protection will need to be considered. In other cases a vehicle may need to be operated in a colder environment, requiring more antifreeze and less water. Three methods are commonly employed to determine the freeze point of the solution:

1. Specific gravity—(using a hydrometer test strip or some sort of floating indicator),

2. Refractometer—which measures the refractive index of the antifreeze solution and translates it into freeze point, and

3. Test strips—specialized, disposable indicators made for this purpose.

Although ethylene glycol hydrometers are widely available and mass-marketed for antifreeze testing, they give false readings at high temperatures because specific gravity changes with temperature. Propylene glycol solutions cannot be tested using specific gravity because of ambiguous results (40% and 100% solutions have the same specific gravity).

Corrosion Inhibitors

Most commercial antifreeze formulations include corrosion inhibiting compounds, and a colored dye (commonly a fluorescent green, red, orange, yellow, or blue) to aid in identification. A 1:1 dilution with water is usually used, resulting in a freezing point of about −34 °F (−37 °C), depending on the formulation. In warmer or colder areas, weaker or stronger dilutions are used, respectively, but a range of 40%/60% to 60%/40% is frequently specified to ensure corrosion protection, and 70%/30% for maximum freeze prevention down to −84 °F (−64 °C).

Maintenance

In the absence of leaks, antifreeze chemicals such as ethylene glycol or propylene glycol may retain their basic properties indefinitely. By contrast, corrosion inhibitors are gradually used up, and must be replenished from time to time. Larger systems (such as HVAC systems) are often monitored by specialist firms which take responsibility for adding corrosion inhibitors and regulating coolant composition. For simplicity, most automotive manufacturers recommend periodic complete replacement of engine coolant, to simultaneously renew corrosion inhibitors and remove accumulated contaminants.

Traditional Inhibitors

Traditionally, there were two major corrosion inhibitors used in vehicles: silicates and phosphates. American made vehicles traditionally used both silicates and phosphates. European makes contain silicates and other inhibitors, but no phosphates. Japanese makes traditionally use phosphates and other inhibitors, but no silicates.

Organic Acid Technology

Certain cars are built with organic acid technology (OAT) antifreeze (e.g., DEX-COOL), or with a hybrid organic acid technology (HOAT) formulation (e.g., Zerex G-05), both of which are claimed to have an extended service life of five years or 240,000 km (150,000 mi).

DEX-COOL specifically has caused controversy. Litigation has linked it with intake manifold gasket failures in General Motors' (GM's) 3.1L and 3.4L engines, and with other failures in 3.8L and 4.3L engines. One of the anti-corrosion components presented as sodium or Potassium 2-ethylhexanoate and ethylhexanoic acid is incompatible with nylon 6,6 and silicone rubber, and is a known plasticizer. Class action lawsuits were registered in several states, and in Canada, to address some of these claims. The first of these to reach a decision was in Missouri, where a settlement was announced early in December 2007. Late in March 2008, GM agreed to compensate complainants in the remaining 49 states. GM (Motors Liquidation Company) filed for bankruptcy in 2009, which tied up the outstanding claims until a court determines who gets paid.

According to the DEX-COOL manufacturer, "mixing a 'green' [non-OAT] coolant with DEX-COOL reduces the batch's change interval to 2 years or 30,000 miles, but will otherwise cause no damage to the engine". DEX-COOL antifreeze uses two inhibitors: sebacate and 2-EHA (2-ethylhexanoic acid), the latter which works well with the hard water found in the United States, but is a plasticizer that can cause gaskets to leak.

According to internal GM documents, the ultimate culprit appears to be operating vehicles for long periods of time with low coolant levels. The low coolant is caused by pressure caps that fail in the open position. (The new caps and recovery bottles were introduced at the same time as DEX-COOL). This exposes hot engine components to air and vapors, causing corrosion and contamination of the coolant with iron oxide particles, which in turn can aggravate the pressure cap problem as contamination holds the caps open permanently.

Honda and Toyota's new extended life coolant use OAT with sebacate, but without the 2-EHA. Some added phosphates provide protection while the OAT builds up. Honda specifically excludes 2-EHA from their formulas.

Typically, OAT antifreeze contains an orange dye to differentiate it from the conventional glycol-based coolants (green or yellow). Some of the newer OAT coolants claim to be compatible with *all* types of OAT and glycol-based coolants; these are typically green or yellow in color.

Hybrid Organic Acid Technology

HOAT coolants typically mix an OAT with a traditional inhibitor, such as silicates or phosphates.

G05 is a low-silicate, phosphate free formula that includes the benzoate inhibitor.

Additives

All automotive antifreeze formulations, including the newer organic acid (OAT antifreeze) formulations, are environmentally hazardous because of the blend of additives (around 5%), including lubricants, buffers and corrosion inhibitors. Because the additives in antifreeze are proprietary, the safety data sheets (SDS) provided by the manufacturer list only those compounds which are considered to be significant safety hazards when used in accordance with the manufacturer's recommendations. Common additives include sodium silicate, disodium phosphate, sodium molybdate, sodium borate, denatonium benzoate and dextrin (hydroxyethyl starch). Disodium fluorescein dyes are added to antifreeze to help trace the source of leaks, and as an identifier since some different formulations are incompatible.

Automotive antifreeze has a characteristic odor due to the additive tolytriazole, a corrosion inhibitor. The unpleasant odor in industrial use tolytriazole comes from impurities in the product that are formed from the toluidine isomers (ortho-, meta- and para-toluidine) and meta-diamino toluene which are side-products in the manufacture of tolytriazole. These side-products are highly reactive and produce volatile aromatic amines which are responsible for the unpleasant odor.

Turbocharger

A turbocharger, colloquially known as a turbo, is a turbine-driven forced induction device that increases an internal combustion engine's efficiency and power output by forcing extra compressed air into the combustion chamber. This improvement over a naturally aspirated engine's power output is due to the fact that the compressor can force more air—and proportionately more fuel—into the combustion chamber than atmospheric pressure (and for that matter, ram air intakes) alone.

Cut-away view of an air foil bearing-supported turbocharger.

Turbochargers were originally known as turbosuperchargers when all forced induction devices were classified as superchargers. Today the term "supercharger" is typically applied only to mechanically driven forced induction devices. The key difference between a turbocharger and a conventional supercharger is that a supercharger is mechanically driven by the engine, often through a belt connected to the crankshaft, whereas a turbocharger is powered by a turbine driven by the engine's exhaust gas. Compared with a mechanically driven supercharger, turbochargers tend to be more efficient, but less responsive. Twincharger refers to an engine with both a supercharger and a turbocharger.

Manufacturers commonly use turbochargers in truck, car, train, aircraft, and construction-equipment engines. They are most often used with Otto cycle and Diesel cycle internal combustion engines.

Forced induction dates from the late 19th century, when Gottlieb Daimler patented the technique of using a gear-driven pump to force air into an internal combustion engine in 1885. The turbocharger was invented by Swiss engineer Alfred Büchi, the head of diesel engine research at Gebrüder Sulzer (now simply called Sulzer), engine manufacturing company in Winterthur, who received a patent in 1905 for using a compressor driven by exhaust gases to force air into an internal combustion engine to increase power output, but it took another 20 years for the idea to come to fruition. The first use of turbocharging technology based on his design was for large marine engines, when the German Ministry of Transport commissioned the construction of the "Preussen" and "Hansestadt Danzig" passenger liners in 1923. Both ships featured twin ten-cylinder diesel engines with output boosted from 1,300 to 1,860 kilowatts (1,750 to 2,500 hp) by turbochargers designed by Büchi and built under his supervision by Brown Boveri (BBC) (now ABB). During World War I French engineer Auguste Rateau fitted turbochargers to Renault engines powering various French fighters with some success. In 1918, General Electric engineer Sanford Alexander Moss attached a turbocharger to a V12 *Liberty* aircraft engine. The engine was tested at Pikes Peak in Colorado at 14,000 ft (4,300 m) to demonstrate that it could eliminate the power loss usually experienced in internal combustion engines as a result of reduced air pressure and density at high altitude.

Turbochargers were first used in production aircraft engines such as the Napier Lioness in the 1920s, although they were less common than engine-driven centrifugal superchargers. Ships and locomotives equipped with turbocharged diesel engines began appearing in the 1920s. Turbochargers were also used in aviation, most widely used by the United States. During World War II, notable examples of U.S. aircraft with turbochargers—which included mass-produced ones designed by General Electric for American aviation use—include the B-17 Flying Fortress, B-24 Liberator, P-38 Lightning, and P-47 Thunderbolt. The technology was also used in experimental fittings by a number of other manufacturers, notably a variety of experimental inline engine-powered Focke-Wulf Fw 190 prototype models, with some developments for their design coming from the DVL, a predecessor of today's DLR agency, but the need for advanced high-temperature metals in the turbine, that were not readily available for production purposes during wartime, kept them out of widespread use.

Turbochargers are widely used in car and commercial vehicles because they allow smaller-capacity engines to have improved fuel economy, reduced emissions, higher power and considerably higher torque.

Turbocharging versus Supercharging

In contrast to turbochargers, superchargers are mechanically driven by the engine. Belts, chains, shafts, and gears are common methods of powering a supercharger, placing a mechanical load on the engine. For example, on the single-stage single-speed supercharged Rolls-Royce Merlin engine, the supercharger uses about 150 hp (110 kW). Yet the benefits outweigh the costs; for the 150 hp (110 kW) to drive the supercharger the engine generates an additional 400 hp (300 kW), a net gain of 250 hp (190 kW). This is where the principal disadvantage of a supercharger becomes apparent; the engine must withstand the net power output of the engine plus the power to drive the supercharger.

Another disadvantage of some superchargers is lower adiabatic efficiency when compared with turbochargers (especially Roots-type superchargers). Adiabatic efficiency is a measure of a compressor's ability to compress air without adding excess heat to that air. Even under ideal conditions, the compression process always results in elevated output temperature; however, more efficient compressors produce less excess heat. Roots superchargers impart significantly more heat to the air than turbochargers. Thus, for a given volume and pressure of air, the turbocharged air is cooler, and as a result denser, containing more oxygen molecules, and therefore more potential power than the supercharged air. In practical application the disparity between the two can be dramatic, with turbochargers often producing 15% to 30% more power based solely on the differences in adiabatic efficiency (however, due to heat transfer from the hot exhaust, considerable heating does occur).

By comparison, a turbocharger does not place a direct mechanical load on the engine, although turbochargers place exhaust back pressure on engines, increasing pumping losses. This is more efficient because while the increased back pressure taxes the piston exhaust stroke, much of the energy driving the turbine is provided by the still-expanding exhaust gas that would otherwise be wasted as heat through the tailpipe. In contrast to supercharging, the primary disadvantage of turbocharging is what is referred to as "lag" or "spool time". This is the time between the demand for an increase in power (the throttle being opened) and the turbocharger(s) providing increased intake pressure, and hence increased power.

Throttle lag occurs because turbochargers rely on the buildup of exhaust gas pressure to drive the turbine. In variable output systems such as automobile engines, exhaust gas pressure at idle, low engine speeds, or low throttle is usually insufficient to drive the turbine. Only when the engine reaches sufficient speed does the turbine section start to *spool up,* or spin fast enough to produce intake pressure above atmospheric pressure.

A combination of an exhaust-driven turbocharger and an engine-driven supercharger can mitigate the weaknesses of both. This technique is called twincharging.

In the case of Electro-Motive Diesel's two-stroke engines, the mechanically assisted turbocharger is not specifically a twincharger, as the engine uses the mechanical assistance to charge air only at lower engine speeds and startup. Once above notch # 5, the engine uses true turbocharging. This differs from a turbocharger that uses the compressor section of the turbo-compressor only during starting and, as a two-stroke engines cannot naturally aspirate, and, according to SAE definitions, a two-stroke engine with a mechanically assisted compressor during idle and low throttle is considered naturally aspirated.

Operating Principle

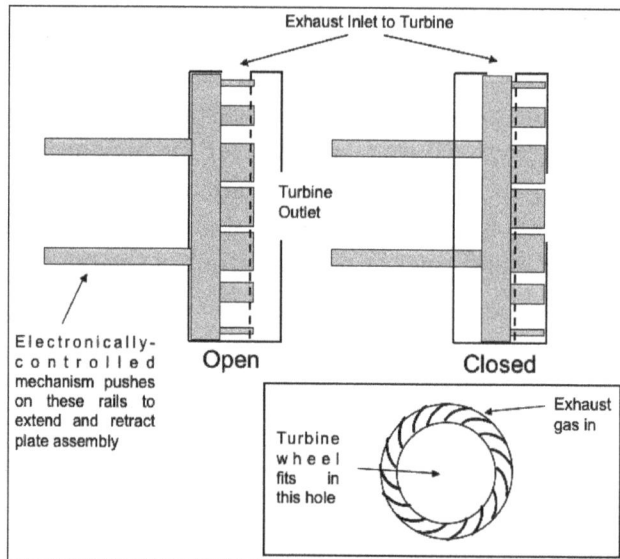

In naturally aspirated piston engines, intake gases are drawn or "pushed" into the engine by atmospheric pressure filling the volumetric void caused by the downward stroke of the piston (which creates a low-pressure area), similar to drawing liquid using a syringe. The amount of air actually inspired, compared with the theoretical amount if the engine could maintain atmospheric pressure, is called volumetric efficiency. The objective of a turbocharger is to improve an engine's volumetric efficiency by increasing density of the intake gas (usually air) allowing more power per engine cycle.

The turbocharger's compressor draws in ambient air and compresses it before it enters into the intake manifold at increased pressure. This results in a greater mass of air entering the cylinders on each intake stroke. The power needed to spin the centrifugal compressor is derived from the kinetic energy of the engine's exhaust gases.

In automotive applications, 'boost' refers to the amount by which intake manifold pressure exceeds atmospheric pressure. This is representative of the extra air pressure that is achieved over what would be achieved without the forced induction. The level of boost may be shown on a pressure gauge, usually in bar, psi or possibly kPa. The control of turbocharger boost has changed dramatically over the 100-plus years of their use. Modern turbochargers can use wastegates, blow-off valves and variable geometry.

In petrol engine turbocharger applications, boost pressure is limited to keep the entire engine system, including the turbocharger, inside its thermal and mechanical design operating range. Over-boosting an engine frequently causes damage to the engine in a variety of ways including pre-ignition, overheating, and over-stressing the engine's internal hardware. For example, to avoid engine knocking (also known as detonation) and the related physical damage to the engine, the intake manifold pressure must not get too high, thus the pressure at the intake manifold of the engine must be controlled by some means. Opening the wastegate allows the excess energy destined for the turbine to bypass it and pass directly to the exhaust pipe, thus reducing boost pressure. The wastegate can be either controlled manually (frequently seen in aircraft) or by an actuator (in automotive applications, it is often controlled by the engine control unit).

Pressure Increase (or Boost)

A turbocharger may also be used to increase fuel efficiency without increasing power. This is achieved by diverting exhaust waste energy, from the combustion process, and feeding it back into the turbo's "hot" intake side that spins the turbine. As the hot turbine side is being driven by the exhaust energy, the cold intake turbine (the other side of the turbo) compresses fresh intake air and drives it into the engine's intake. By using this otherwise wasted energy to increase the mass of air, it becomes easier to ensure that all fuel is burned before being vented at the start of the exhaust stage. The increased temperature from the higher pressure gives a higher Carnot efficiency.

A reduced density of intake air is caused by the loss of atmospheric density seen with elevated altitudes. Thus, a natural use of the turbocharger is with aircraft engines. As an aircraft climbs to higher altitudes, the pressure of the surrounding air quickly falls off. At 18,000 feet (5,500 m), the air is at half the pressure of sea level, which means that the engine produces less than half-power at this altitude. In aircraft engines, turbocharging is commonly used to maintain manifold pressure as altitude increases (i.e. to compensate for lower-density air at higher altitudes). Since atmospheric pressure reduces as the aircraft climbs, power drops as a function of altitude in normally aspirated engines. Systems that use a turbocharger to maintain an engine's sea-level power output are called turbo-normalized systems. Generally, a turbo-normalized system attempts to maintain a manifold pressure of 29.5 inHg (100 kPa).

Turbocharger Lag

Turbocharger lag (turbo lag) is the time required to change power output in response to a throttle change, noticed as a hesitation or slowed *throttle response* when accelerating as compared to a naturally aspirated engine. This is due to the time needed for the exhaust system and turbocharger to generate the required boost which can also be referred to as spooling. Inertia, friction, and compressor load are the primary contributors to turbocharger lag. Superchargers do not suffer this problem, because the turbine is eliminated due to the compressor being directly powered by the engine.

Turbocharger applications can be categorized into those that require changes in output power (such as automotive) and those that do not (such as marine, aircraft, commercial automotive, industrial, engine-generators, and locomotives). While important to varying degrees, turbocharger lag is most problematic in applications that require rapid changes in power output. Engine designs reduce lag in a number of ways:

- Lowering the rotational inertia of the turbocharger by using lower radius parts and ceramic and other lighter materials.

- Changing the turbine's *aspect ratio*.

- Increasing upper-deck air pressure (compressor discharge) and improving wastegate response.

- Reducing bearing frictional losses, e.g., using a foil bearing rather than a conventional oil bearing.

- Using variable-nozzle or twin-scroll turbochargers.

- Decreasing the volume of the upper-deck piping.

- Using multiple turbochargers sequentially or in parallel.

- Using an antilag system.

- Using a turbocharger spool valve to increase exhaust gas flow speed to the (twin-scroll) turbine.

Sometimes turbo lag is mistaken for engine speeds that are below boost threshold. If engine speed is below a turbocharger's boost threshold rpm then the time needed for the vehicle to build speed and rpm could be considerable, maybe even tens of seconds for a heavy vehicle starting at low vehicle speed in a high gear. This wait for vehicle speed increase is not turbo lag, it is improper gear selection for boost demand. Once the vehicle reaches sufficient speed to provide the required rpm to reach boost threshold, there will be a far shorter delay while the turbo itself builds rotational energy and transitions to positive boost, only this last part of the delay in achieving positive boost is the turbo lag.

Boost Threshold

The *boost threshold* of a turbocharger system is the lower bound of the region within which the compressor operates. Below a certain rate of flow, a compressor produces insignificant boost. This limits boost at a particular RPM, regardless of exhaust gas pressure. Newer turbocharger and engine developments have steadily reduced boost thresholds.

Electrical boosting ("E-boosting") is a new technology under development. It uses an electric motor to bring the turbocharger up to operating speed quicker than possible using available exhaust gases. An alternative to e-boosting is to completely separate the turbine and compressor into a turbine-generator and electric-compressor as in the hybrid turbocharger. This makes compressor speed independent of turbine speed. In 1981, a similar system that used a hydraulic drive system and overspeed clutch arrangement accelerated the turbocharger of the MV *Canadian Pioneer*.

Turbochargers start producing boost only when a certain amount of kinetic energy is present in the exhaust gasses. Without adequate exhaust gas flow to spin the turbine blades, the turbocharger cannot produce the necessary force needed to compress the air going into the engine. The boost threshold is determined by the engine displacement, engine rpm, throttle opening, and the size of the turbocharger. The operating speed (rpm) at which there is enough exhaust gas momentum to compress the air going into the engine is called the "boost threshold rpm". Reducing the "boost threshold rpm" can improve throttle response.

Key Components

The turbocharger has three main components:

- The turbine, which is almost always a radial inflow turbine (but is almost always a single-stage axial inflow turbine in large Diesel engines).

- The compressor, which is almost always a centrifugal compressor.

- The center housing/hub rotating assembly.

Many turbocharger installations use additional technologies, such as wastegates, intercooling and blow-off valves.

Turbine

On the left, the brass oil drain connection. On the right are the braided oil supply line and water coolant line connections.

Compressor impeller side with the cover removed.

Energy provided for the turbine work is converted from the enthalpy and kinetic energy of the gas. The turbine housings direct the gas flow through the turbine as it spins at up to 250,000 rpm. The size and shape can dictate some performance characteristics of the overall turbocharger. Often the same basic turbocharger assembly is available from the manufacturer with multiple housing choices for the turbine, and sometimes the compressor cover as well. This lets the balance between performance, response, and efficiency be tailored to the application.

The turbine and impeller wheel sizes also dictate the amount of air or exhaust that can flow through the system, and the relative efficiency at which they operate. In general, the larger the turbine wheel and compressor wheel the larger the flow capacity. Measurements and shapes can vary, as well as curvature and number of blades on the wheels.

Turbine side housing removed.

A turbocharger's performance is closely tied to its size. Large turbochargers take more heat and pressure to spin the turbine, creating lag at low speed. Small turbochargers spin quickly, but may not have the same performance at high acceleration. To efficiently combine the benefits of large and small wheels, advanced schemes are used such as twin-turbochargers, twin-scroll turbochargers, or variable-geometry turbochargers.

Twin-turbo

Twin-turbo or bi-turbo designs have two separate turbochargers operating in either a sequence or in parallel. In a parallel configuration, both turbochargers are fed one-half of the engine's exhaust. In a sequential setup one turbocharger runs at low speeds and the second turns on at a predetermined engine speed or load. Sequential turbochargers further reduce turbo lag, but require an intricate set of pipes to properly feed both turbochargers.

Two-stage variable twin-turbos employ a small turbocharger at low speeds and a large one at higher speeds. They are connected in a series so that boost pressure from one turbocharger is multiplied by another, hence the name "2-stage." The distribution of exhaust gas is continuously variable, so the transition from using the small turbocharger to the large one can be done incrementally. Twin turbochargers are primarily used in Diesel engines. For example, in Opel bi-turbo Diesel, only the smaller turbocharger works at low speed, providing high torque at 1,500–1,700 rpm. Both turbochargers operate together in mid range, with the larger one pre-compressing the air, which the smaller one further compresses. A bypass valve regulates the exhaust flow to each turbocharger. At higher speed (2,500 to 3,000 RPM) only the larger turbocharger runs.

Smaller turbochargers have less turbo lag than larger ones, so often two small turbochargers are used instead of one large one. This configuration is popular in engines over 2,5 litres and in V-shape or boxer engines.

Twin-scroll

Twin-scroll or divided turbochargers have two exhaust gas inlets and two nozzles, a smaller sharper angled one for quick response and a larger less angled one for peak performance.

With high-performance camshaft timing, exhaust valves in different cylinders can be open at the same time, overlapping at the end of the power stroke in one cylinder and the end of exhaust stroke in another. In twin-scroll designs, the exhaust manifold physically separates the channels for cylinders that can interfere with each other, so that the pulsating exhaust gasses flow through separate spirals (scrolls). With common firing order 1–3–4–2, two scrolls of unequal length pair cylinders 1 and 4, and 3 and 2. This lets the engine efficiently use exhaust scavenging techniques, which decreases exhaust gas temperatures and NO x emissions, improves turbine efficiency, and reduces turbo lag evident at low engine speeds.

Cut-out of a twin-scroll turbocharger, with two differently angled nozzles.

Cut-out of a twin-scroll exhaust and turbine; the dual "scrolls" pairing cylinders 1 and 4, and 2 and 3 are clearly visible.

Variable-geometry

Garrett variable-geometry turbocharger on DV6TED4 engine.

Variable-geometry or variable-nozzle turbochargers use moveable vanes to adjust the air-flow to the turbine, imitating a turbocharger of the optimal size throughout the power curve. The vanes are placed just in front of the turbine like a set of slightly overlapping walls. Their angle is adjusted by an actuator to block or increase air flow to the turbine. This variability maintains a comparable exhaust velocity and back pressure throughout the engine's rev range. The result is that the turbocharger improves fuel efficiency without a noticeable level of turbocharger lag.

Compressor

The compressor increases the mass of intake air entering the combustion chamber. The compressor is made up of an impeller, a diffuser and a volute housing.

The operating range of a compressor is described by the "compressor map".

Ported Shroud

The flow range of a turbocharger compressor can be increased by allowing air to bleed from a ring of holes or a circular groove around the compressor at a point slightly downstream of the compressor inlet (but far nearer to the inlet than to the outlet).

The ported shroud is a performance enhancement that allows the compressor to operate at significantly lower flows. It achieves this by forcing a simulation of impeller stall to occur continuously. Allowing some air to escape at this location inhibits the onset of surge and widens the operating range. While peak efficiencies may decrease, high efficiency may be achieved over a greater range of engine speeds. Increases in compressor efficiency result in slightly cooler (more dense) intake air, which improves power. This is a passive structure that is constantly open (in contrast to compressor exhaust blow off valves, which are mechanically or electronically controlled). The ability of the compressor to provide high boost at low rpm may also be increased marginally (because near choke conditions the compressor draws air inward through the bleed path). Ported shrouds are used by many turbocharger manufacturers.

Center Housing/Hub Rotating Assembly

The center hub rotating assembly (CHRA) houses the shaft that connects the compressor impeller and turbine. It also must contain a bearing system to suspend the shaft, allowing it to rotate at very high speed with minimal friction. For instance, in automotive applications the CHRA typically uses a thrust bearing or ball bearing lubricated by a constant supply of pressurized engine oil. The CHRA may also be considered "water-cooled" by having an entry and exit point for engine coolant. Water-cooled models use engine coolant to keep lubricating oil cooler, avoiding possible oil coking (destructive distillation of engine oil) from the extreme heat in the turbine. The development of air-foil bearings removed this risk.

Ball bearings designed to support high speeds and temperatures are sometimes used instead of fluid bearings to support the turbine shaft. This helps the turbocharger accelerate more quickly and reduces turbo lag. Some variable nozzle turbochargers use a rotary electric actuator, which uses a direct stepper motor to open and close the vanes, rather than pneumatic controllers that operate based on air pressure.

Additional Technologies Commonly used in Turbocharger Installations

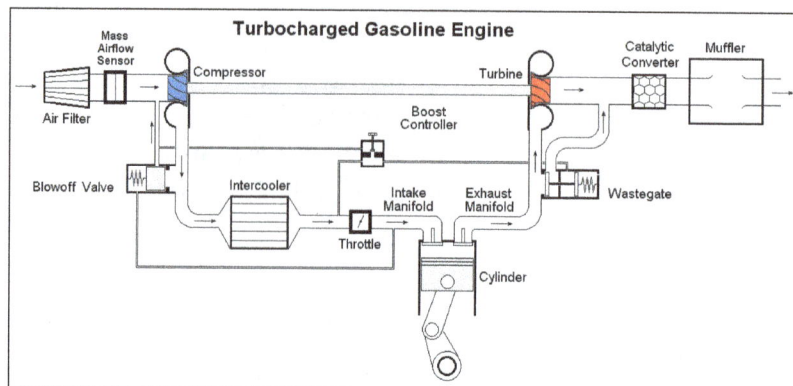

Illustration of typical component layout in a production turbocharged gasoline engine.

Intercooling

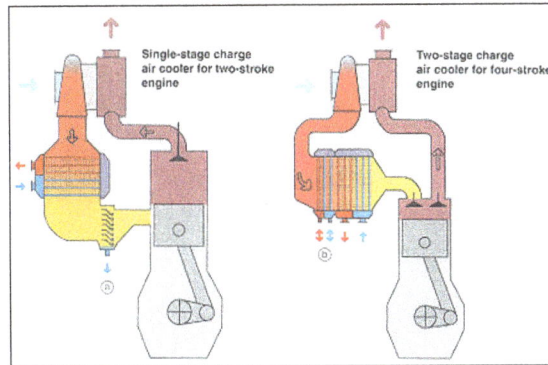

Illustration of inter-cooler location.

When the pressure of the engine's intake air is increased, its temperature also increases. This occurrence can be explained through Gay-Lussac's law, stating that the pressure of a given amount of gas held at constant volume is directly proportional to the Kelvin temperature. With more pressure being added to the engine through the turbocharger, overall temperatures of the engine will also rise. In addition, heat soak from the hot exhaust gases spinning the turbine will also heat the intake air. The warmer the intake air, the less dense, and the less oxygen available for the combustion event, which reduces volumetric efficiency. Not only does excessive intake-air temperature reduce efficiency, it also leads to engine knock, or detonation, which is destructive to engines.

To compensate for the increase in temperature, turbocharger units often make use of an intercooler between successive stages of boost to cool down the intake air. A *charge air cooler* is an air cooler between the boost stage(s) and the appliance that consumes the boosted air.

Top-mount (TMIC) vs. Front-mount Intercoolers (FMIC)

There are two areas on which intercoolers are commonly mounted. It can be either mounted on top, parallel to the engine, or mounted near the lower front of the vehicle. Top-mount intercoolers setups will result in a decrease in turbo lag, due in part by the location of the intercooler being much closer to the turbocharger outlet and throttle body. This closer proximity reduces the time it takes for air to travel through the system, producing power sooner, compared to that of a front-mount intercooler which has more distance for the air to travel to reach the outlet and throttle.

Front-mount intercoolers can have the potential to give better cooling compared to that of a top-mount. The area in which a top-mounted intercooler is located, is near one of the hottest areas of a car, right above the engine. This is why most manufacturers include large hood scoops to help feed air to the intercooler while the car is moving, but while idle, the hood scoop provides little to no benefit. Even while moving, when the atmospheric temperatures begin to rise, top-mount intercoolers tend to underperform compared to front-mount intercoolers. With more distance to travel, the air circulated through a front-mount intercooler may have more time to cool.

Water Injection

An alternative to intercooling is injecting water into the intake air to reduce the temperature. This method has been used in automotive and aircraft applications.

Methanol Injection

Methanol/water injection has been around since the 1920s but was not utilized until World War II. Adding the mixture to intake of the turbocharged engines decreased operating temperatures and increased horse power. Turbocharged engines today run high boost and high engine temperatures to match. When injecting the mixture into the intake stream, the air is cooled as the liquids evaporate. Inside the combustion chamber it slows the flame, acting similar to higher octane fuel. Methanol/water mixture allows for higher compression because of the less detonation-prone and, thus, safer combustion inside the engine.

Fuel-air Mixture Ratio

In addition to the use of intercoolers, it is common practice to add extra fuel to the intake air (known as "running an engine rich") for the sole purpose of cooling. The amount of extra fuel varies, but typically reduces the air-fuel ratio to between 11 and 13, instead of the stoichiometric 14.7 (in petrol engines). The extra fuel is not burned (as there is insufficient oxygen to complete the chemical reaction), instead it undergoes a phase change from atomized (liquid) to gas. This phase change absorbs heat, and the added mass of the extra fuel reduces the average thermal energy of the charge and exhaust gas. Even when a catalytic converter is used, the practice of running an engine rich increases exhaust emissions.

Wastegate

A wastegate regulates the exhaust gas flow that enters the exhaust-side driving turbine and therefore the air intake into the manifold and the degree of boosting. It can be controlled by a boost pressure assisted, generally vacuum hose attachment point diaphragm (for vacuum and positive pressure to return commonly oil contaminated waste to the emissions system) to force the spring-loaded diaphragm to stay closed until the overboost point is sensed by the ecu or a solenoid operated by the engine's electronic control unit or a boost controller, but most production vehicles use a single vacuum hose attachment point spring-loaded diaphragm that can alone be pushed open, thus limiting overboost ability due to exhaust gas pressure forcing open the wastegate.

Anti-Surge/Dump/Blow off Valves

A recirculating type anti-surge valve.

Turbocharged engines operating at wide open throttle and high rpm require a large volume of air to flow between the turbocharger and the inlet of the engine. When the throttle is closed, compressed air flows to the throttle valve without an exit (i.e., the air has nowhere to go).

In this situation, the surge can raise the pressure of the air to a level that can cause damage. This is because if the pressure rises high enough, a compressor stall occurs—stored pressurized air decompresses backward across the impeller and out the inlet. The reverse flow back across the turbocharger makes the turbine shaft reduce in speed more quickly than it would naturally, possibly damaging the turbocharger.

To prevent this from happening, a valve is fitted between the turbocharger and inlet, which vents off the excess air pressure. These are known as an anti-surge, diverter, bypass, turbo-relief valve, blow-off valve (BOV), or dump valve. It is a pressure relief valve, and is normally operated by the vacuum from the intake manifold.

The primary use of this valve is to maintain the spinning of the turbocharger at a high speed. The air is usually recycled back into the turbocharger inlet (diverter or bypass valves), but can also be vented to the atmosphere (blow off valve). Recycling back into the turbocharger inlet is required on an engine that uses a mass-airflow fuel injection system, because dumping the excessive air overboard downstream of the mass airflow sensor causes an excessively rich fuel mixture—because the mass-airflow sensor has already accounted for the extra air that is no longer being used. Valves that recycle the air also shorten the time needed to re-spool the turbocharger after sudden engine deceleration, since load on the turbocharger when the valve is active is much lower than if the air charge vents to atmosphere.

Free Floating

A free floating turbocharger is the simplest type of turbocharger. This configuration has no wastegate and cannot control its own boost levels. They are typically designed to attain maximum boost at full throttle. Free floating turbochargers produce more horsepower because they have less backpressure, but are not driveable in performance applications without an external wastegate.

A free floating turbocharger is used in the 100 litre engine of this Caterpillar mining vehicle.

Applications

Petrol-powered Cars

The first turbocharged passenger car was the Oldsmobile Jetfire option on the 1962–1963 F85/

Cutlass, which used a turbocharger mounted to a 215 cu in (3.52 L) all aluminum V8. Also in 1962, Chevrolet introduced a special run of turbocharged Corvairs, initially called the Monza Spyder and later renamed the Corsa, which mounted a turbocharger to its air cooled flat six cylinder engine. This model popularized the turbocharger in North America—and set the stage for later turbocharged models from Porsche on the 1975-up 911/930, Saab on the 1978–1984 Saab 99 Turbo, and the very popular 1978–1987 Buick Regal/T Type/Grand National. Today, turbocharging is common on both diesel and gasoline-powered cars. Turbocharging can increase power output for a given capacity or increase fuel efficiency by allowing a smaller displacement engine. The 'Engine of the year 2011' is an engine used in a Fiat 500 equipped with an MHI turbocharger. This engine lost 10% weight, saving up to 30% in fuel consumption while delivering the same HP (105) as a 1.4 litre engine.

Diesel-powered Cars

The first production turbocharger diesel passenger car was the Garrett-turbocharged Mercedes 300SD introduced in 1978. Today, most automotive diesels are turbocharged, since the use of turbocharging improved efficiency, driveability and performance of diesel engines, greatly increasing their popularity. The Audi R10 with a diesel engine even won the 24 hours race of Le Mans in 2006, 2007 and 2008.

Motorcycles

The first example of a turbocharged bike is the 1978 Kawasaki Z1R TC. Several Japanese companies produced turbocharged high-performance motorcycles in the early 1980s, such as the CX500 Turbo from Honda- a transversely mounted, liquid cooled V-Twin also available in naturally aspirated form. Since then, few turbocharged motorcycles have been produced. This is partially due to an abundance of larger displacement, naturally aspirated engines being available that offer the torque and power benefits of a smaller displacement engine with turbocharger, but do return more linear power characteristics. The Dutch manufacturer EVA motorcycles builds a small series of turbocharged diesel motorcycle with an 800cc smart CDI engine.

Trucks

The first turbocharged diesel truck was produced by *Schweizer Maschinenfabrik Saurer* (Swiss Machine Works Saurer) in 1938.

Aircraft

A natural use of the turbocharger—and its earliest known use for any internal combustion engine, starting with experimental installations in the 1920s—is with aircraft engines. As an aircraft climbs to higher altitudes the pressure of the surrounding air quickly falls off. At 5,486 m (18,000 ft), the air is at half the pressure of sea level and the airframe experiences only half the aerodynamic drag. However, since the charge in the cylinders is pushed in by this air pressure, the engine normally produces only half-power at full throttle at this altitude. Pilots would like to take advantage of the low drag at high altitudes to go faster, but a naturally aspirated engine does not produce enough power at the same altitude to do so.

The table below is used to demonstrate the wide range of conditions experienced. As seen in the table below, there is significant scope for forced induction to compensate for lower density environments.

	Daytona Beach	Denver	Death Valley	Colorado State Highway 5	La Rinconada, Peru.
elevation	0 m / 0 ft	1,609 m / 5,280 ft	−86 m / −282 ft	4,347 m / 14,264 ft	5,100 m / 16,732 ft
atm	1.000	0.823	1.010	0.581	0.526
bar	1.013	0.834	1.024	0.589	0.533
psia	14.696	12.100	14.846	8.543	7.731
kPa	101.3	83.40	102.4	58.90	53.30

A turbocharger remedies this problem by compressing the air back to sea-level pressures (turbo-normalizing), or even much higher (turbo-charging), in order to produce rated power at high altitude. Since the size of the turbocharger is chosen to produce a given amount of pressure at high altitude, the turbocharger is oversized for low altitude. The speed of the turbocharger is controlled by a wastegate. Early systems used a fixed wastegate, resulting in a turbocharger that functioned much like a supercharger. Later systems utilized an adjustable wastegate, controlled either manually by the pilot or by an automatic hydraulic or electric system. When the aircraft is at low altitude the wastegate is usually fully open, venting all the exhaust gases overboard. As the aircraft climbs and the air density drops, the wastegate must continuously close in small increments to maintain full power. The altitude at which the wastegate fully closes and the engine still produces full power is the *critical altitude*. When the aircraft climbs above the critical altitude, engine power output decreases as altitude increases, just as it would in a naturally aspirated engine.

With older supercharged aircraft without Automatic Boost Control, the pilot must continually adjust the throttle to maintain the required manifold pressure during ascent or descent. The pilot must also take care to avoid over-boosting the engine and causing damage. In contrast, modern turbocharger systems use an automatic wastegate, which controls the manifold pressure within parameters preset by the manufacturer. For these systems, as long as the control system is working properly and the pilot's control commands are smooth and deliberate, a turbocharger cannot over-boost the engine and damage it.

Yet the majority of World War II engines used superchargers, because they maintained three significant manufacturing advantages over turbochargers, which were larger, involved extra piping, and required exotic high-temperature materials in the turbine and pre-turbine section of the exhaust system. The size of the piping alone is a serious issue; American fighters Vought F4U and Republic P-47 used the same engine, but the huge barrel-like fuselage of the latter was, in part, needed to hold the piping to and from the turbocharger in the rear of the plane. Turbocharged piston engines are also subject to many of the same operating restrictions as gas turbine engines. Pilots must make smooth, slow throttle adjustments to avoid overshooting their target manifold pressure. The fuel/air mixture must often be adjusted far on the rich side of stoichiometric combustion needs to avoid pre-ignition or detonation in the engine when running at high power settings. In systems using a manually operated wastegate, the pilot must be careful not to exceed the turbocharger's maximum rpm. The additional systems and piping increase an aircraft engine's size, weight, complexity and cost. A turbocharged aircraft engine costs more to maintain than a comparable normally aspirated engine. The great majority of World War II American heavy bombers used by the USAAF, particularly the Wright R-1820 *Cyclone-9* powered B-17 Flying Fortress, and Pratt & Whitney R-1830 Twin Wasp powered Consolidated B-24 Liberator four-engine bombers both used similar models of General Electric-designed turbochargers in service, as did the twin Allison V-1710-engined Lockheed P-38 Lightning American heavy fighter during the war years.

All of the above WWII aircraft engines had mechanically driven centrifugal superchargers as-designed from the start, and the turbosuperchargers (with intercoolers) were added, effectively as twincharger systems, to achieve desired altitude performance.

Today, most general aviation piston engine powered aircraft are naturally aspirated. Modern aviation piston engines designed to run at high altitudes typically include a turbocharger (either high pressure or turbonormalized) rather than a supercharger. The change in thinking is largely due to economics. Aviation gasoline was once plentiful and cheap, favoring the simple, but fuel-hungry, supercharger. As the cost of fuel has increased, the supercharger has fallen out of favor.

Turbocharged aircraft often occupy a performance range between that of normally aspirated piston-powered aircraft and turbine-powered aircraft. Despite the negative points, turbocharged aircraft fly higher for greater efficiency. High cruise flight also allows more time to evaluate issues before a forced landing must be made.

As the turbocharged aircraft climbs, however, the pilot (or automated system) can close the wastegate, forcing more exhaust gas through the turbocharger turbine, thereby maintaining manifold pressure during the climb, at least until the critical pressure altitude is reached (when the wastegate is fully closed), after which manifold pressure falls. With such systems, modern high-performance piston engine aircraft can cruise at altitudes up to 25,000 feet (above which, RVSM certification would be required), where low air density results in lower drag and higher true airspeeds. This allows flying "above the weather". In manually controlled wastegate systems, the pilot must take care not to overboost the engine, which causes detonation, leading to engine damage.

Marine and Land-based Diesel Turbochargers

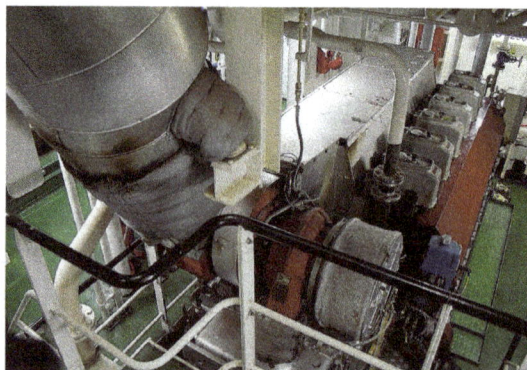

A medium-sized six-cylinder marine diesel-engine, with turbocharger and exhaust in the foreground.

Turbocharging, which is common on diesel engines in automobiles, trucks, tractors, and boats is also common in heavy machinery such as locomotives, ships, and auxiliary power generation.

- Turbocharging can dramatically improve an engine's specific power and power-to-weight ratio, performance characteristics that are normally poor in non-turbocharged diesel engines.

- Diesel engines have no detonation because diesel fuel is injected at or towards the end of the compression stroke and is ignited solely by the heat of compression of the charge air. Because of this, diesel engines can use a much higher boost pressure than spark ignition engines, limited only by the engine's ability to withstand the additional heat and pressure.

Turbochargers are also employed in certain two-stroke cycle diesel engines, which would normally require a Roots blower for aspiration. In this specific application, mainly Electro-Motive Diesel (EMD) 567, 645, and 710 Series engines, the turbocharger is initially driven by the engine's crankshaft through a gear train and an overrunning clutch, thereby providing aspiration for combustion. After combustion has been achieved, and after the exhaust gases have reached sufficient heat energy, the overrunning clutch is automatically disengaged, and the turbo-compressor is thereafter driven exclusively by the exhaust gases. In the EMD application, the turbocharger acts as a compressor for normal aspiration during starting and low power output settings and is used for true turbocharging during medium and high power output settings. This is particularly beneficial at high altitudes, as are often encountered on western U.S. railroads. It is possible for the turbocharger to revert to compressor mode momentarily during commands for large increases in engine power.

Boost Controller

A boost controller is a device to control the boost level produced in the intake manifold of a turbocharged or supercharged engine by affecting the air pressure delivered to the pneumatic and mechanical wastegate actuator.

A boost controller can be a simple manual control which can be easily fabricated, or it may be included as part of the engine management computer in a factory turbocharged car, or an aftermarket electronic boost controller.

Principles of Operation

Without a boost controller, air pressure is fed from the charge air (compressed side) of the turbocharger directly to the wastegate actuator via a vacuum hose. This air pressure can come from anywhere on the intake after the turbo, including after the throttle body, though that is less common. This air pressure pushes against the force of a spring located in the wastegate actuator to allow the wastegate to open and re-direct exhaust gas so that it does not reach the turbine wheel. In this simple configuration, the spring's springrate and preload determine how much boost pressure the system will achieve. Springs are classified by the boost pressure they typically achieve, such as a "7 psi spring" that will allow the turbocharger to reach equilibrium at approximately 7 psi (0.48 bar).

One primary problem of this system is the wastegate will start to open well before the actual desired boost pressure is achieved. This negatively affects the threshold of boost onset and also increases turbocharger lag. For instance, a spring rated at 7 psi may allow the wastegate to begin to (but not fully) open at as little as 3.5 psi (0.24 bar).

Achieving moderate boost levels consistently is also troublesome with this configuration. At partial throttle, full boost may still be reached, making the vehicle difficult to control with precision. Electronic systems can allow the throttle to control the level of boost, so that only at full throttle will maximum boost levels be achieved and intermediate levels of boost can be held consistently at partial throttle levels.

Also to be noted is the way in which boost control is achieved, depending on the type of wastegate used. Typically manual "bleed type" boost controllers are only used on swing type (single port) wastegate actuators. To increase boost, pressure is taken away from the actuator control line, therefore increasing the turbo output pressure required to counteract the controllers leak-lowered pressure acting on the wastegate. Dual port swing type wastegate actuators and external wastegates generally require electronic boost control although adjustable boost control can also be achieved on both of these with an air pressure regulator, this is not the same as a bleed type boost controller. To increase boost with an external or dual port wastegate, pressure is added to the top control port to increase boost. When boost control is not fitted, this control port is open to the atmosphere.

Manual Boost Control

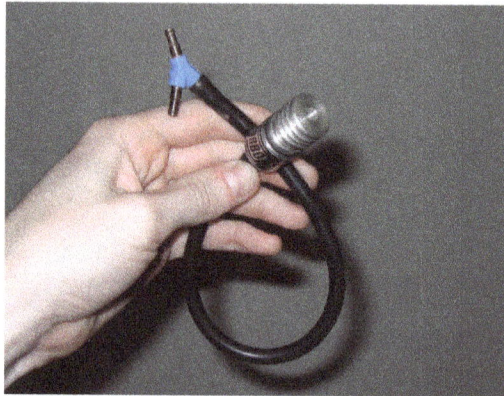

A simple manual boost controller. A small screw is located in the top of the aluminum body to adjust bleed rate. This model is placed in the engine bay; however, the vacuum line could be extended to allow it to reach into the passenger compartment.

A bleed-type manual boost controller simple mechanical and pneumatic control to allow some pressure from the wastegate actuator to escape or bleed out to the atmosphere or back into the intake system. This can be as simple as a T-fitting on the boost control line near the actuator with a small bleeder screw. The screw can be turned out to varying degrees to allow air to bleed out of the system, relieving pressure on the wastegate actuator, thus increasing boost levels. These devices are popular due to their negligible cost compared to other devices that may offer the same power increase.

A ball & spring type boost controller uses the force of a spring acting against the boost pressure to control boost. This is installed with one boost signal line coming from the intake somewhere after the turbocharger, and one boost signal line going to the wastegate. A knob changes the force on the spring which in turn dictates how much pressure is on the ball. The tighter the spring, the more boost that is needed to unseat the ball, and allow the boost pressure to reach the wastegate actuator. There is a bleed hole on the boost controller after the ball, to allow the pressurized air that would be trapped between the wastegate actuator and the ball after it is seated again. These type of Manual boost controllers are very popular since they do not provide a boost leak, allowing faster spool times and better control than a "bleed type" boost controller.

There are several different designs of ball-and-spring controllers on the market that range greatly in terms of cost and quality. Common body materials are brass and aluminum vary from inline to 90 degree designs. Another design aspect is the ball valve seat which is critical for performance stability.

Generally a manual boost controller will not be located within the cabin of the vehicle as the lengthy vacuum piping run between the turbo/wastegate & controller can introduce response issues into the system. It is possible to use two manual boost controllers at different settings with a solenoid to switch between them for two different boost pressure settings. Some factory turbocharged cars have a switch to regulate boost pressure, such as a setting designed for fuel economy and a setting for performance.

Manual boost controllers cannot be used to set a specific boost level at a given throttle position (& therefore be used to optimise driveability & control issues), although a ball-spring type boost controller does allow the boost threshold to be as low as is possible on a given engine configuration, and also keeps turbo spool as fast as is possible as the wastegate remains completely shut until the desired boost pressure is reached, ensuring 100% of the exhaust gases are diverted through the turbocharger exhaust turbine. They can be used in conjunction with *some* electronic systems.

Electronic Boost Control

Electronic boost control adds an air control solenoid and/or a stepper motor controlled by an electronic control unit. The same general principle of a manual controller is present, which is to control the air pressure presented to the wastegate actuator. Further control and intelligent algorithms can be introduced, refining and increasing control over actual boost pressure delivered to the engine.

At the component level, boost pressure can either be bled out of the control lines or blocked outright. Either can achieve the goal of reducing pressure pushing against the wastegate. In a bleed-type system air is allowed to pass out of the control lines, reducing the load on the wastegate actuator. On a blocking configuration, air traveling from the charge air supply to the wastegate actuator is blocked while simultaneously bleeding any pressure that has previously built up at the wastegate actuator.

A 3-port pneumatic solenoid. This solenoid allows interrupt or blocking of the boost pressure rather than just bleed type control.

Control Details

A 4-port pneumatic solenoid installed to control a dual port
wastegate controlled by a single PWM PID controller.

Control for the solenoids and stepper motors can be either closed loop or open loop. Closed loop systems rely on feedback from a manifold pressure sensor to meet a predetermined boost pressure. Open loop systems have a predetermined control output where control output is merely based on other inputs such as throttle angle and/or engine RPM. Open loop specifically leaves out a desired boost level, while closed loop attempts to target a specific level of boost pressure. Since open loop systems do not modify control levels based on MAP sensor, differing boost pressure levels may be reached based on outside variables such as weather conditions or engine coolant temperature. For this reason, systems that do not feature closed loop operation are not as widespread.

Boost controllers often use pulse width modulation (PWM) techniques to bleed off boost pressure on its way to the reference port on the wastegate actuator diaphragm in order to (on occasion) under report boost pressure in such a way that the wastegate permits a turbocharger to build more boost pressure in the intake than it normally could. In effect, a boost-control solenoid valve lies to the wastegate under the engine control unit´s (ECU) control. The boost control solenoid contains a needle valve that can open and close very quickly. By varying the pulse width to the solenoid, the solenoid valve can be commanded to be open a certain percentage of the time. This effectively alters the flow rate of air pressure through the valve, changing the rate at which air bleeds out of the T in the manifold pressure reference line to the wastegate. This effectively changes the air pressure as seen by the wastegate actuator diaphragm. Solenoids may require small diameter restrictors be installed in the air control lines to limit airflow and even out the on/off nature of their operation.

The wastegate control solenoid can be commanded to run in a variety of frequencies in various gears, engine speeds, or according to various other factors in a deterministic open-loop mode. Or, by monitoring manifold pressure in a feedback loop, the engine management system can monitor the efficacy of PWM changes in the boost control solenoid bleed rate at altering boost pressure in the intake manifold, increasing or decreasing the bleed rate to target a particular maximum boost.

The basic algorithm sometimes involves the EMS (engine management system) "learning" how quickly the turbocharger can spool and how quickly the boost pressure increases. Armed with this knowledge, as long as boost pressure is below a predetermined allowable ceiling, the EMS will open the boost control solenoid to allow the turbocharger to create overboost beyond what the wastegate would normally allow. As overboost reaches the programmable maximum, the EMS begins to decrease the bleed rate through the control solenoid to raise boost pressure as seen at the wastegate actuator diaphragm so the wastegate opens enough to limit boost to the maximum configured level of over-boost.

Stepper motors allow fine control of airflow based on position and speed of the motor, but may have low total airflow capability. Some systems use a solenoid in conjunction with a stepper motor, with the stepper motor allowing fine control and the solenoid coarse control.

Many configurations are possible with 2, 3, and 4 port solenoids and stepper motors in series or parallel. Two-port solenoid bleed systems with a PID controller tend to be common on factory turbocharged cars.

Advantages

Since less positive pressure can be present at the wastegate actuator as desired boost is approached the wastegate remains closer to a completely closed state. This keeps exhaust gas routed through the turbine and increases energy transferred to the wheels of the turbocharger. Once desired boost is reached, closed loop based systems react by allowing more air pressure to reach the wastegate actuator to stop the further increase in air pressure so desired boost levels are maintained. This reduces turbocharger lag and lowers boost threshold. Boost pressure builds faster when the throttle is depressed quickly and allows boost pressure to build at lower engine RPM than without such a system.

This also allows the use of a much softer spring in the actuator. For instance, a 7 psi (0.48 bar) spring together with a boost controller may still be able to achieve a maximum boost level of well over 15 psi (1.0 bar). The electronic control unit can be programmed to control 7 psi (0.48 bar) psi at half throttle, 12 psi (0.83 bar) at 3/4 throttle, and 15 psi (1.0 bar) at full throttle, or whatever levels the programmer or designer of the control unit intends. This partial throttle control greatly increases driver control over the engine and vehicle.

Limitations and Disadvantages

Even with an electronic controller, actuator springs that are too soft can cause the wastegate to open before desired. Exhaust gas backpressure is still pushing against the wastegate valve itself. This backpressure can overcome the spring pressure without the aid of the actuator at all. Electronic control may still enable control of boost to over double gauge pressure of the spring's rated pressure.

The solenoid and stepper motors also need to be installed in such a way to maximize the advantages of failure modes. For instance, if a solenoid is installed to control boost electronically, it should be installed such that if the solenoid fails in the most common failure mode (probably non-energized position) the boost control falls back to simple wastegate actuator boost levels. It is possible a solenoid or stepper motor could get stuck in a position that lets no boost pressure reach the wastegate, causing boost to quickly rise out of control.

The electronic systems, extra hoses, solenoids and soforth add complexity to the turbocharger system. This runs counter to the "keep it simple" principle as there are more things that can go wrong.

It is worth noting that virtually all modern factory turbocharged cars, the same cars with long warranty periods, implement electronic boost control. Manufacturers such as Subaru, Mitsubishi and Saab integrate electronic boost control in all turbo model cars.

Availability and Applications

Electronic boost control systems are available as aftermarket stand-alone systems such as the HKS EVC and VBC, Apex-i AVC-R, GFB G-force, or Gizzmo IBC / MS-IBC as a built-in feature of modern factory turbocharged vehicles such as the Subaru Impreza WRX STi and often as built-in features in full aftermarket stand-alone engine management systems such as the Holley EFI, Hydra Nemesis, AEM EMS and MegaSquirt.

Dangers in use

Installing a boost controller in a vehicle that is already well tuned, such as a factory turbocharged car, may allow higher boost pressure than tolerable by the engine or turbocharger, reducing life and reliability. Care should be taken to avoid exceeding the limits of any engine system components such as the engine block, fuel injectors, or engine management system. This is as true with boost control as it is with fuel and timing controls, or any number of other engine system modifications.

In particular, users may find the extremely low cost and ease of adding a manual boost controller a particular draw for extra power at low cost compared to more comprehensive modifications. Users should carefully consider how installing any boost controller may affect and interact with existing complex engine management systems. Additional boost levels may not be tolerated by the existing turbocharger, causing faster wear. Fuel injectors or the fuel pump may not be able to deliver additional fuel needed for higher air flow and power of higher boost pressure. Or the engine management system may not be able to properly compensate for fuel or ignition timing, causing knock and/or engine failure.

Past and Future

There are other outdated methods of boost control, such as intake restriction or bleed off. For instance, it is possible to install a large butterfly valve in the intake to restrict airflow as desired boost is approached. It is also possible to actually release large amounts of already compressed air similar to a blowoff valve but on a constant basis to maintain desired boost at the intake manifold. The currently popular exhaust gas bypass via wastegate is quite superior if compared to creating intake restriction or wasting energy by releasing air that has already been compressed. These methods are rarely used in modern system due to the large sacrifices in efficiency, heat, and reliability.

Other methods may come into widespread use in the future, such as variable geometry turbochargers. With a sufficiently large turbine, no wastegate is necessary. Low speed response and faster spool up are then obtained using variable turbine technologies rather than a smaller turbine. These systems may replace or supplement typical wastegates as they develop. Control methods for the variable mechanical controls, such as the principles of closed loop will still apply even if they no longer involve pneumatics.

Tuned Exhaust

Ferrari V10 engine showing one of its two tuned extractor manifolds.

A tuned exhaust is an exhaust system for an internal combustion engine which improves its efficiency by using precise geometry to reflect the pressure waves from the exhaust valve or port back to the valve or port at a particular time in the cycle.

Two-stroke Engines

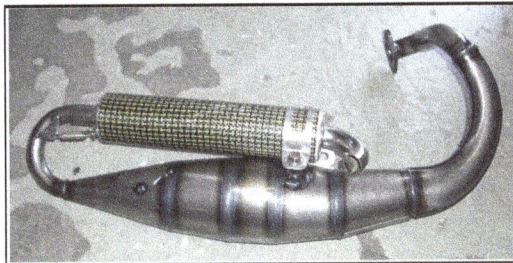

Yasuni aftermarket motor scooter exhaust system. The exhaust passes first through the expansion chamber at the bottom and then exits through the muffler above it.

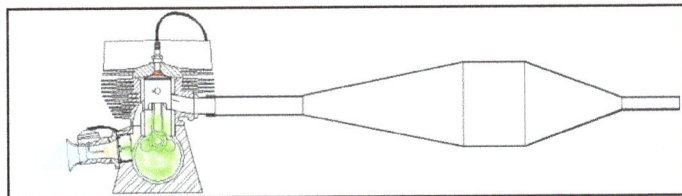

A conceptual animation of a two-stroke engine with a tuned exhaust system using an expansion chamber. Exhaust gases are in grey, fuel/air mixture is green. In practice the fuel/air mix is unlikely to progress as far down the exhaust pipe as shown.

In many two-stroke engines, the exhaust port is opened and closed directly by the position of the piston rather than by a separate valve, which restricts the timing of its operation; Typically, the port remains open long after is optimum, allowing some of the incoming charge to escape. This can be partly addressed by use of a tuned exhaust system to deliver a pulse of positive pressure prior to the port closing, to retain the charge.

Alternatives

Direct-injection two-stroke diesel engines tend to use exhaust valves actuated either by camshafts or electronic control, rather than exhaust ports. This system is called uniflow scavenging. Opposed piston engines are inherently uniflow-scavenged, but these do use piston-controlled cylinder ports. Two-stroke opposed piston engines such as the Napier Deltic and Junkers Jumo 204 engines use one piston to control the inlet port and the other the exhaust, allowing more flexibility in timing. A variation of this approach is taken by the split-single engine, in which two cylinders share one combustion chamber, with the piston in one cylinder controlling the transfer port and the other the exhaust port.

Four-stroke Engines

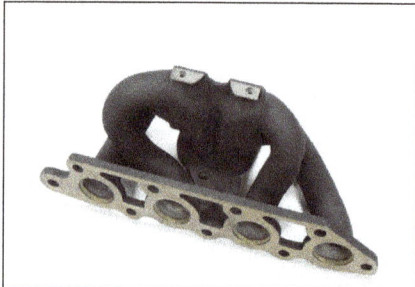

Aftermarket extractor manifold.

Rotax 912s aero engine showing the tuned exhaust system.

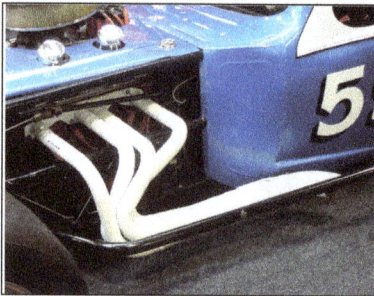

Collector on a racing car.

Zoomie headers on a dragster.

Extractor Manifolds

Most non-turbo performance cars and high-performance four-stroke motorcycles use extractor manifolds (headers in American English), as do most non-turbo racing cars. Extractor manifolds are also available as aftermarket accessories to suit many engines.

Extractor manifolds offer the following advantages over the simple manifolds often fitted to non-performance engines:

- Separating the gas flows from the individual cylinders so that undesirable inter-cylinder interference is avoided.

- Maintaining an optimum gas velocity by carefully chosen tube diameter.

- Allowing the individual cylinders to assist one another by means of the negative pressure waves generated at the *collector*, where the individual exhausts merge.

This type of exhaust system can be used with or without a muffler, and so can be used on both race and road vehicles.

Dual Mass Flywheel

The Dual Mass Flywheel, or DMF, is a major technological advancement in automotive transmission systems. The push for sustainability in the transport sector has driven the hybrid and electric revolution, but even conventional petrol-powered vehicles have seen significant fuel economy advancements, resulting in lower overall CO_2 emissions.

A major part of these advancements is the direct result of smaller engines. Auto makers are now designing cars with three and even two-cylinder engines, and while these smaller engines have been successful in reducing fuel consumption, they are now being asked to provide the torque and power of much bigger motors. The result is a significant increase in vibration and noise, particularly at low speeds.

A DMF acts in much the same way as a traditional, single flywheel – they provide direct contact between the engine and the clutch assembly in manual transmissions. Where DMFs differ from single flywheels is more than the fact that there are two flywheels as opposed to one – it's what happens between the two flywheels that makes all the difference.

DMFs incorporate a series of springs between the flywheels and these springs act as vibration dampeners. Where vibration and noise in transmissions using a single flywheel has nowhere to go except directly into the powertrain system, a DMF's spring system dampens this engine vibration, resulting in less noise, increased comfort for the driver, and increased life of the transmission.

What's the downside? Having a DMF installed has generally meant higher maintenance costs for drivers. Clutch replacement has often required DMF replacement at the same time, and DMFs have traditionally been expensive and time consuming for workshops to install.

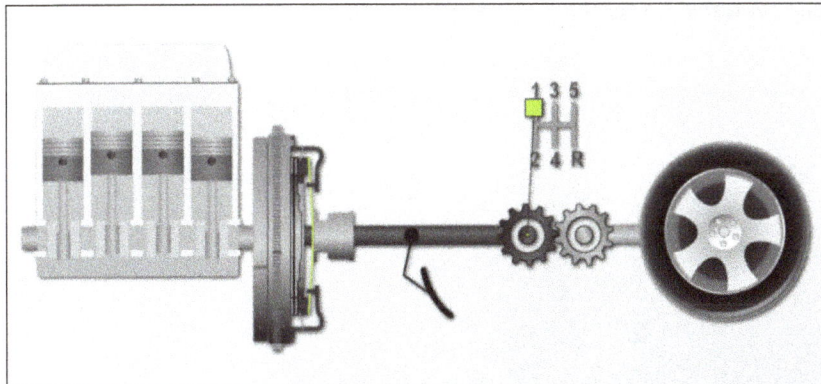

But some recent developments are making replacing DMFs much more affordable.

Major DMF manufacturers like Valeo are producing DMFs that are simple for mechanics to install and require no special tools, meaning significantly reduced fitting costs.

Valeo have also recently introduced a new type of DMF, the VBlade. Instead of a series of springs between the two flywheels, the VBlade utilises two vibration dampening blades. The result is an incredibly durable DMF that will mean lower maintenance costs for drivers.

Valeo VBlade DMF.

Dual Mass Flywheels have proven to be so successful in reducing vibration and noise that now one in every two vehicles rolling off the assembly line is fitted with a DMFs.

And as manufacturers like Valeo continue to develop DMF technology – increasing their durability and reducing their cost – DMFs look like they will be a major part of automotive powertrain systems for years to come.

References

- "Using an Engine Heater in a Diesel Engine for Cold-Weather Starts". www.dummies.com. Retrieved 27 August 2019

- dual-mass-flywheel: theengineer.co.uk, Retrieved 14 April, 2019

- Goodsell D (December 2009). "Molecule of the Month: Antifreeze Proteins". The Scripps Research Institute and the RCSB PDB. doi:10.2210/rcsb_pdb/mom_2009_12

- "Exhaust System Technology: Science and Implementation of High Performance Exhaust Systems". www.epi-eng.com. Retrieved 25 October 2019

- Evaluation of Certain Food Additives and Contaminants (Technical Report Series). World Health Organization. p. 105. ISBN 92-4-120909-7

- "The turbocharger turns 100 years old this week". www.newatlas.com. 18 November 2005. Retrieved 20 September 2019

Permissions

Index